W9-BLQ-865

UNKNOWN QUANTITY

UNKNOWN QUANTITY

A REAL AND IMAGINARY HISTORY OF ALGEBRA

John Derbyshire

Joseph Henry Press
Washington, D.C.

Joseph Henry Press • 500 Fifth Street, NW • Washington, DC 20001

The Joseph Henry Press, an imprint of the National Academies Press, was created with the goal of making books on science, technology, and health more widely available to professionals and the public. Joseph Henry was one of the early founders of the National Academy of Sciences and a leader in early American science.

Any opinions, findings, conclusions, or recommendations expressed in this volume are those of the author and do not necessarily reflect the views of the National Academy of Sciences or its affiliated institutions.

Library of Congress Cataloging-in-Publication Data

Derbyshire, John.
  The unknown quantity : a real and imaginary history of algebra / by John Derbyshire.
      p. cm.
  Includes bibliographical references and index.
  ISBN 0-309-09657-X (hardback : alk. paper) 1. Algebra—History. 2. Equations—History. 3. Algebra, Universal—History. 4. Algebra, Abstract—History. 5. Geometry, Algebraic—History. I. Title.
  QA151.D47 2006
  512.009—dc22
                                              2005037018

Cover image: J-L Charmet / Photo Researchers, Inc.

Printed in the United States of America.

For Rosie

CONTENTS

Part 3
Levels of Abstraction

INTRODUCTION

§1.1 THIS BOOK IS A HISTORY OF ALGEBRA, written for the curious nonmathematician. It seems to me that the author of such a book should begin by telling his reader what algebra *is*. So what is it?

Passing by an airport bookstore recently, I spotted a display of those handy crib sheets used by high school and college students, the ones that have all the basics of a subject printed on a folding triptych laminated in clear plastic. There were two of these cribs for algebra, titled "Algebra—Part 1" and "Algebra—Part 2." Parts 1 and 2 combined (said the subheading) "cover principles for basic, intermediate, and college courses."[1]

I read through the material they contained. Some of the topics might not be considered properly algebraic by a professional mathematician. "Functions," for example, and "Sequences and Series" belong to what professional mathematicians call "analysis." On the whole, though, this is a pretty good summary of basic algebra and reveals the working definition of the word "algebra" in the modern American high school and college-basics curriculum: Algebra is the part of advanced mathematics that is not calculus.

In the higher levels of math, however, algebra has a distinctive quality that sets it apart as a discipline by itself. One of the most famous quotations in 20th-century math is this one, by the great German mathematician Hermann Weyl. It appeared in an article he published in 1939.

> In these days the angel of topology and the devil of abstract algebra
> fight for the soul of each individual mathematical domain.[2]

Perhaps the reader knows that topology is a branch of geometry, sometimes called "rubber-sheet geometry," dealing with the properties of figures that are unchanged by stretching and squeezing, but not cutting, the figure. (The reader who does not know this will find a fuller description of topology in §14.2. And for more on the context of Weyl's remark, see §14.6.) Topology tells us the difference between a plain loop of string and one that is knotted, between the surface of a sphere and the surface of a doughnut. Why did Weyl place these harmless geometrical investigations in such a strong opposition to algebra?

Or look at the list of topics in §15.1, mentioned in citations for the Frank Nelson Cole Prize in Algebra during recent years. Unramified class field theory . . . Jacobean variety . . . function fields . . . motivitic cohomology. . . . Plainly we are a long way removed here from quadratic equations and graphing. What is the common thread? The short answer is hinted at in that quote from Hermann Weyl: It is *abstraction*.

§I.2   All of mathematics is abstract, of course. The very first act of mathematical abstraction occurred several millennia ago when human beings discovered numbers, taking the imaginative leap from observed instances of (for example) "three-ness"—three fingers, three cows, three siblings, three stars—to *three*, a mental object that could be contemplated by itself, without reference to any particular instance of three-ness.

The second such act, the rise to a second level of abstraction, was the adoption, in the decades around 1600 CE, of literal symbolism— that is, the use of letter symbols to represent arbitrary or unknown numbers: *data* (things given) or *quaesita* (things sought). "Universal arithmetic," Sir Isaac Newton called it. The long, stumbling journey to this point had been motivated mainly by the desire to solve equations, to determine the *unknown quantity* in some mathematical situation. It was that journey, described in Part 1 of my book, that planted the word "algebra" in our collective consciousness.

A well-educated person of the year 1800 would have said, if asked, that algebra was just that—the use of letter symbols to "relieve the imagination" (Leibniz) when carrying out arithmetic and solving equations. By that time the mastery of, or at least some acquaintance with, the use of literal symbolism for math was part of a general European education.

During the 19th century[3] though, these letter symbols began to detach themselves from the realm of numbers. Strange new mathematical objects[4] were discovered[5]: groups, matrices, manifolds, and many others. Mathematics began to soar up to new levels of abstraction. That process was a natural development of the use of literal symbolism, once that symbolism had been thoroughly internalized by everyone. It is therefore not unreasonable to regard it as a continuation of the history of algebra.

I have accordingly divided my narrative into three parts, as follows:

> *Part 1:* From the earliest times to the adoption of a systematic literal symbolism—letters representing numbers—around the year 1600.

> *Part 2:* The first mathematical victories of that symbolism and the slow detachment of symbols from the concepts of traditional arithmetic and geometry, leading to the discovery of new mathematical objects.

Part 3: Modern algebra—the placing of the new mathematical objects on a firm logical foundation and the ascent to ever higher levels of abstraction.

Because the development of algebra was irregular and haphazard, in the way of all human things, I found it difficult to keep to a strictly chronological approach, especially through the 19th century. I hope that my narrative makes sense nonetheless and that the reader will get a clear view of all the main lines of development.

§I.3   My aim is not to teach higher algebra to the reader. There are plenty of excellent textbooks for that: I shall recommend some as I go along. This book is not a textbook. I hope only to show what algebraic ideas are *like*, how the later ones developed from the earlier ones, and what kind of people were responsible for it all, in what kind of historical circumstances.

   I did find it impossible, though, to describe the history of this subject without some minimal explanation of what these algebraists were *doing*. There is consequently a fair amount of math in this book. Where I have felt the need to go beyond what is normally covered in high school courses, I have "set up" this material in brief math primers here and there throughout the text. Each of these primers is placed at the point where you will need to read through it in order to continue with the historical narrative. Each provides some basic concepts. In some cases I enlarge on those concepts in the main text; the primers are intended to jog the memory of a reader who has done some undergraduate courses or to provide very basic understanding to a reader who hasn't.

§I.4   This book is, of course, a work of secondary exposition, drawn mostly from other people's books. I shall credit those books in the text and Endnotes as I go along. There are, however, three sources that I refer to so often that I may as well record my debt to them here at the beginning. The first is the invaluable *Dictionary of Scientific Biography*, referred to hereinafter, as *DSB*, which not only provides details of the lives of mathematicians but also gives valuable clues about how mathematical ideas originate and are transmitted.

The other two books I have relied on most heavily are histories of algebra written by mathematicians for mathematicians: *A History of Algebra* by B. L. van der Waerden (1985) and *The Beginnings and Evolution of Algebra* by Isabella Bashmakova and Galina Smirnova (translated by Abe Shenitzer, 2000). I shall refer to these books in what follows just by the names of their authors ("van der Waerden says . . .").

One other major credit belongs here. I had the great good fortune to have my manuscript looked over at a late stage in its development by Professor Richard G. Swan of the University of Chicago. Professor Swan offered numerous comments, criticisms, corrections, and suggestions, which together have made this a better book than it would otherwise have been. I am profoundly grateful to him for his help and encouragement. "Better" is not "perfect," of course, and any errors or omissions that still lurk in these pages are entirely my own responsibility.

§I.5 Here, then, is the story of algebra. It all began in the remote past, with a simple turn of thought from the declarative to the interrogative, from "this plus this equals this" to "this plus *what* equals this?" The unknown quantity—the x that everyone associates with algebra—first entered human thought right there, dragging behind it, at some distance, the need for a symbolism to represent unknown or arbitrary numbers. That symbolism, once established, allowed the study of equations to be carried out at a higher level of abstraction. As a result, new mathematical objects came to light, leading up to yet higher levels.

In our own time, algebra has become the most rarefied and demanding of all mental disciplines, whose objects are abstractions of abstractions of abstractions, yet whose results have a power and beauty that are all too little known outside the world of professional mathematicians. Most amazing, most mysterious of all, these ethereal mental objects seem to contain, within their nested abstractions, the deepest, most fundamental secrets of the physical world.

Math Primer

NUMBERS AND POLYNOMIALS

§NP.1   AT INTERVALS THROUGH THIS BOOK I shall interrupt the historical narrative with a math primer, giving very brief coverage of some
math topic you need to know, or be reminded of, in order to follow
the history.

This first math primer stands before the entire book. There are
two concepts you need to have a good grasp of in order to follow
anything at all in the main narrative. Those two concepts are *number*
and *polynomial*.

§NP.2   The modern conception of number—it began to take shape
in the late 19th century and became widespread among working
mathematicians in the 1920s and 1930s—is the nested "Russian dolls"
model. There are five Russian dolls in the model, denoted by "hollow
letters" $\mathbb{N}$, $\mathbb{Z}$, $\mathbb{Q}$, $\mathbb{R}$, and $\mathbb{C}$ and remembered by the nonsense mnemonic: "Nine Zulu Queens Ruled China."

The innermost doll is the *natural numbers*, collectively denoted
by the symbol $\mathbb{N}$. These are the ordinary[6] counting numbers: 1, 2,
3, . . . They can be arranged pictorially as a line of dots extending indefinitely to the right:

o   o   o   o   o   o   o   o   o   o   o   · · ·

1   2   3   4   5   6   7   8   9   10   11   · · ·

FIGURE NP-1 The family of natural numbers, ℕ.

The natural numbers are very useful, but they have some short-comings. The main shortcomings are that you can't always subtract one natural number from another or divide one natural number by another. You can subtract 5 from 7, but you can't subtract 12 from 7—not, I mean, if you want a natural-number answer. Term of art: ℕ is not *closed under subtraction.* ℕ is not closed under division either: You can divide 12 by 4 but not by 5, not without falling over the edge of ℕ into some other realm.

The subtraction problem was solved by the discovery of zero and the negative numbers. Zero was discovered by Indian mathematicians around 600 CE. Negative numbers were a fruit of the European Renaissance. Expanding the system of natural numbers to include these new kinds of numbers gives the second Russian doll, enclosing the first one. This is the system of *integers,* collectively denoted by ℤ (from the German word *Zahl,* "number"). The integers can be pictured by a line of dots extending indefinitely to both left and right:

· · ·   o   o   o   o   o   o   o   o   o   · · ·

· · ·   −4   −3   −2   −1   0   1   2   3   4   · · ·

FIGURE NP-2 The family of integers, ℤ.

We can now add, subtract, and multiply at will, though multiplication needs a knowledge of the *rule of signs*:

A positive times a positive gives a positive.
A positive times a negative gives a negative.
A negative times a positive gives a negative.
A negative times a negative gives a positive.

Or more succinctly: Like signs give a positive; unlike signs give a negative. The rule of signs applies to division, too, when it is possible. So −12 divided by −3 gives 4.

Division, however, is not usually possible. $\mathbb{Z}$ is not closed under division. To get a system of numbers that is closed under division, we expand yet again, bringing in the fractions, both positive and negative ones. This makes a third Russian doll, containing both the first two. This doll is called the *rational numbers*, collectively denoted by $\mathbb{Q}$ (from "quotient").

The rational numbers are "dense." This means that between any two of them, you can always find another one. Neither $\mathbb{N}$ nor $\mathbb{Z}$ has this property. There is no natural number to be found between 11 and 12. There is no integer to be found between −107 and −106. There is, however, a rational number to be found between $\frac{1190507}{10292801}$ and $\frac{185015}{1399602}$, even though these two numbers differ by less than 1 part in 16 trillion. The rational number $\frac{2300597}{19890493}$, for example, is greater than the first of those rational numbers, but less than the second. It is easy to show that since there is a rational number between any two rational numbers, you can find as many rational numbers as you please between any two rational numbers. That's the real meaning of "dense."

Because $\mathbb{Q}$ has this property of being dense, it can be illustrated by a continuous line stretching away indefinitely to the left and right. Every rational number has a position on that line.

FIGURE NP-3 The family of rational numbers, $\mathbb{Q}$.
(Note: This same figure serves to illustrate the family of real numbers, $\mathbb{R}$.)

See how the gaps between the integers are all filled up? Between any two integers, say 27 and 28, the rational numbers are dense.

These Russian dolls are nested, remember. $\mathbb{Q}$ includes $\mathbb{Z}$, and $\mathbb{Z}$ includes $\mathbb{N}$. Another way to look at this is: A natural number is an "honorary integer," and an integer—or, for that matter, a natural number—is an "honorary rational number." The honorary number can, for purposes of emphasis, be dressed up in the appropriate costume. The natural number 12 can be dressed up as the integer +12, or as the rational number $\frac{12}{1}$.

§NP.3   That there are other kinds of numbers, neither whole nor rational, was discovered by the Greeks about 500 BCE. The discovery made a profound impression on Greek thought and raised questions that even today have not been answered to the satisfaction of all mathematicians and philosophers.

The simplest example of such a number is the square root of 2—the number that, if you multiply it by itself, gives the answer 2. (Geometrically: The diagonal of a square whose sides are one unit in length.) It is easy to show that no rational number can do this.[7] Very similar arguments show that if N is not a perfect kth power, the kth root of N is not rational.

Plainly we need another Russian doll to encompass all these irrationals. This new doll is the system of *real numbers*, denoted in the aggregate by $\mathbb{R}$. The square root of 2 is a real number but not a rational number: It is in $\mathbb{R}$ but not in $\mathbb{Q}$ (let alone $\mathbb{Z}$ or $\mathbb{N}$, of course).

The real numbers, like the rational numbers, are dense. Between any two of them, you can always find another one. Since the rational numbers are already dense—already "fill up" the illustrative line—you might wonder how the real numbers can be squinched in among them. The whole business is made even stranger by the fact that $\mathbb{Z}$ and $\mathbb{Q}$ are "countable," but $\mathbb{R}$ is not. A countable set is a set you can match off with the counting numbers $\mathbb{N}$: one, two, three, . . . , even if the tally needs to go on forever. You can't do that with $\mathbb{R}$. There is a sense in which $\mathbb{R}$ is "too big" to tally like that—bigger than $\mathbb{N}$, $\mathbb{Z}$, and $\mathbb{Q}$. So however can this superinfinity of real numbers be fitted in among the rational numbers?

That is a very interesting problem, which has caused mathematicians much vexation. It does not belong in a history of algebra, though, and I mention it here only because there are a couple of passing references to countability later in the book (§14.3 and §14.4). Suffice it to say here that a diagram to illustrate $\mathbb{R}$ looks exactly like the one I just offered for $\mathbb{Q}$: a single continuous line stretching away forever to the left and right (Figure NP-3). When this line is being used to illustrate $\mathbb{R}$, it is called "the real line." More abstractly, "the real line" can be taken as just a synonym for $\mathbb{R}$.

§NP.4   Within $\mathbb{N}$ we could add always, subtract sometimes, multiply always, and divide sometimes. Within $\mathbb{Z}$ we could add, subtract, and multiply always but divide only sometimes. Within $\mathbb{Q}$ we could add, subtract, multiply, and divide at will (except that division by zero is never allowed in math), but extracting roots threw up problems.

$\mathbb{R}$ solved those problems but only for positive numbers. By the rule of signs, any number, when multiplied by itself, gives a positive answer. To say it the other way around: Negative numbers have no square roots in $\mathbb{R}$.

From the 16th century onward this limitation began to be a hindrance to mathematicians, so a new Russian doll was added to the scheme. This doll is the system of *complex numbers*, denoted by $\mathbb{C}$. In it *every* number has a square root. It turns out that you can build up this entire system using just ordinary real numbers, together with one single new number: the number $\sqrt{-1}$, always denoted by i. The square root of –25, for example, is $5i$, because $5i \times 5i = 25 \times (-1)$, which is –25. What about the square root of i? No problem. The familiar rule for multiplying out parentheses is $(u + v) \times (x + y) = ux + uy + vx + vy$. So

$$\left(\frac{1}{\sqrt{2}} + \frac{1}{\sqrt{2}}i\right) \times \left(\frac{1}{\sqrt{2}} + \frac{1}{\sqrt{2}}i\right) = \frac{1}{2} + \frac{1}{2}i + \frac{1}{2}i + \frac{1}{2}i^2,$$

and since $i^2 = -1$ and $\frac{1}{2} + \frac{1}{2} = 1$, that right-hand side is just equal to i. Each of those parentheses on the left is therefore a square root of i.

As before, the Russian dolls are nested. A real number x is an honorary complex number: the complex number $x + 0i$. (A complex number of the form $0 + yi$, or just yi for short, y understood to be a real number, is called an *imaginary number*.)

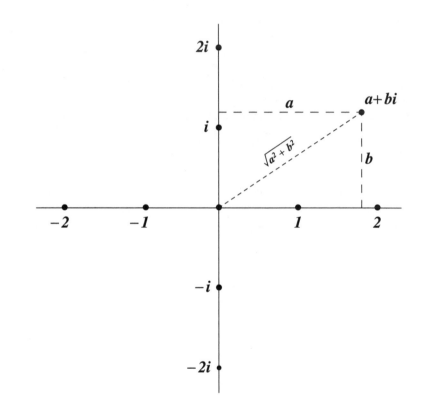

FIGURE NP-4 The family of complex numbers, $\mathbb{C}$.

The rules for adding, subtracting, multiplying, and dividing complex numbers all follow easily from the fact that $i^2 = -1$. Here they are:

Addition: $(a + bi) + (c + di) = (a + c) + (b + d)i$

Subtraction: $(a + bi) - (c + di) = (a - c) + (b - d)i$

Multiplication: $(a + bi) \times (c + di) = (ac - bd) + (ad + bc)i$

Division: $(a + bi) \div (c + di) = \dfrac{ac + bd}{c^2 + d^2} + \dfrac{bc - ad}{c^2 + d^2} i$

Because a complex number has two independent parts, $\mathbb{C}$ can't be illustrated by a line. You need a flat *plane*, going to infinity in all directions, to illustrate $\mathbb{C}$. This is called *the complex plane* (Figure NP-4). A complex number $a + bi$ is represented by a point on the plane, using ordinary west-east, south-north coordinates.

Notice that associated with any complex number $a + bi$, there is a very important positive real number called its *modulus*, defined to be $\sqrt{a^2 + b^2}$. I hope it is plain from Figure NP-4 that, by Pythagoras's theorem,[8] the modulus of a complex number is just its distance from the zero point—always called the *origin*—in the complex plane.

We shall meet some other number systems later, but everything starts from these four basic systems, each nested inside the next: $\mathbb{N}$, $\mathbb{Z}$, $\mathbb{Q}$, $\mathbb{R}$, and $\mathbb{C}$.

§NP.5   So much for numbers. The other key concept I shall refer to freely all through this book is that of a *polynomial*. The etymology of this word is a jumble of Greek and Latin, with the meaning "having many names," where "names" is understood to mean "named parts." It seems to have first been used by the French mathematician François Viète in the late 16th century, showing up in English a hundred years later.

A polynomial is a mathematical *expression* (not an *equation*—there is no equals sign) built up from numbers and "unknowns" by the operations of addition, subtraction, and multiplication only, these operations repeated as many times as you like, though not an infinite number of times. Here are some examples of polynomials:

$$5x^{12} - 22x^7 - 141x^6 + x^3 - 19x^2 - 245$$

$$9x^2 - 13xy + y^2 - 14x - 35y + 18$$

$$2x - 7$$

$$x$$

$$\frac{211}{372}x^4 + \pi x^3 - (7 - 8i)x^2 + \sqrt{3}\,x$$

$$x^2 + x + y^2 + y + z^2 + z + t^2 + t$$

$$ax^2 + bx + c$$

Notice the following things:

Unknowns. There can be any number of unknowns in a polynomial.

Using the alphabet for unknowns. The true unknowns, the ones whose values we are really interested in—Latin *quaesita*, "things sought"—are usually taken from the end of the Latin alphabet: x, y, z, and t are the letters most commonly used for unknowns.

Powers of the unknowns. Since you can do any finite number of multiplications, any natural number power of any unknown can show up: x, x^2, x^3, x^2y^3, x^5yz^2, . . .

Using the alphabet for "givens." The "things given"—Latin *data*—are often just numbers taken from $\mathbb{N}$, $\mathbb{Z}$, $\mathbb{Q}$, $\mathbb{R}$, or $\mathbb{C}$. We may generalize an argument, though, by using letters for the givens. These letters are usually taken either from the beginning of the alphabet (a, b, c, . . .) or from the middle (p, q, r, . . .).

Coefficients. "Data" now has a life of its own as an English word, and hardly anyone says "givens." The "things given" in a polynomial are now called *coefficients.* The coefficients of that third sample polynomial above are 2 and −7. The coefficient of the fourth polynomial (strictly speaking it is a monomial) is 1. The coefficients of the last polynomial are *a*, *b*, and *c*.

§NP.6 Polynomials form just a small subset of all possible mathematical expressions. If you introduce division into the mix, you get a larger class of expressions, called *rational expressions*, like this one:

$$\frac{x^2 - 3y^2}{2xz}$$

which is a rational expression with three unknowns. This is *not* a polynomial. You can enlarge the set further by allowing more operations: the extraction of roots; the taking of sines, cosines, or logarithms; and so on. The expressions you end up with are not polynomials either.

Recipe for a polynomial: Take some "given" numbers, which you may spell out explicitly $(17, \sqrt{2}, \pi, \ldots)$ or hide behind letters from the beginning or middle of the alphabet $(a, b, c, \ldots, p, q, r, \ldots)$. Mix in some unknowns $(x, y, z, \ldots)$. Perform some finite number of additions, subtractions, and multiplications. The result will be a polynomial.

Even though they comprise only a tiny proportion of mathematical expressions, polynomials are tremendously important, especially in algebra. The adjective "algebraic," when used by mathematicians, can usually be translated as "concerned with polynomials." Examine a theorem in algebra, even one at the very highest level. By peeling off a couple of layers of meaning, you will very likely uncover a polynomial. *Polynomial* has a fair claim to being the single most important concept in algebra, both ancient and modern.

Part 1

THE UNKNOWN QUANTITY

Chapter 1

Four Thousand Years Ago

§1.1 IN THE BROAD SENSE I defined in my introduction, the turn of thought from declarative to interrogative arithmetic, algebra began very early in recorded history. Some of the oldest written texts known to us that contain any mathematics at all contain material that can fairly be called algebraic. Those texts date from the first half of the second millennium BCE, from 37 or 38 centuries ago,[9] and were written by people living in Mesopotamia and Egypt.

To a person of our time, that world seems inconceivably remote. The year 1800 BCE was almost as far back in Julius Caesar's past as Caesar is in ours. Outside a small circle of specialists, the only widespread knowledge of that time and those places is the fragmentary and debatable account given in the Book of Genesis and thereby known to all well-instructed adherents of the great Western monotheistic religions. This was the world of Abraham and Isaac, Jacob and Joseph, Ur and Haran, Sodom and Gomorrah. The Western civilization of that time encompassed all of the Fertile Crescent, that nearly continuous zone of cultivable land that stretches northwest from the Persian Gulf up the plains of the Tigris and Euphrates, across the Syrian plateau, and down through Palestine to the Nile delta and Egypt. All the peoples of this zone knew each other. There was con-

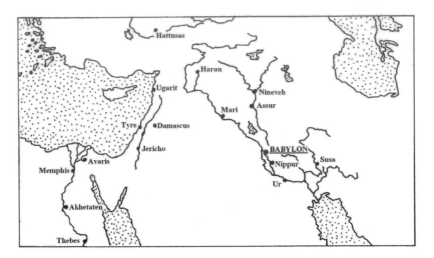

FIGURE 1-1 The Fertile Crescent.

stant traffic all around the Crescent, from Ur on the lower Euphrates
to Thebes on the middle Nile. Abraham's trek from Ur to Palestine,
then to Egypt, would have followed well-traveled roads.

Politically the three main zones of the Fertile Crescent looked
quite different. Palestine was a provincial backwater, a place you went
through to get somewhere else. Peoples of the time regarded it as
within Egypt's sphere of influence. Egypt was ethnically uniform and
had no seriously threatening peoples on her borders. The nation was
a millennium and a half old—older than England is today—before
she suffered her first foreign invasion, of which I shall say more later.
In their self-sufficient security the Egyptians settled early on into a
sort of Chinese mentality, a centralized monarchy ruling through a
vast bureaucratic apparatus recruited by merit. Almost 2,000 official
titles were in use as early as the Fifth Dynasty, around 2500 to 2350
BCE, "so that in the wondrous hierarchy everyone was unequal to
everyone else," as Robert G. Wesson says in *The Imperial Order*.

Mesopotamia presents a different picture. There was much more
ethnic churning, with first Sumerians, then Akkadians, then Elamites,

Amorites, Hittites, Kassites, Assyrians, and Aramaeans ascendant. Egyptian-style bureaucratic despotism sometimes had its hour in Mesopotamia, when a powerful ruler could master enough territory, but these imperial episodes rarely lasted long. The first and most important of them had been Sargon the Great's Akkadian dynasty, which ruled all of Mesopotamia for 160 years, from 2340 to 2180 BCE, before disintegrating under assault by Caucasian tribes. By the time of which I am writing, the 18th and 17th centuries BCE, the Sargonid glory was a fading memory. It had, however, bequeathed to the region a more or less common language: Akkadian, of the Semitic family. Sumerian persisted in the south and apparently also as a sort of prestige language known by educated people, rather like Greek among the Romans or Latin in medieval and early-modern Europe.

The normal condition of Mesopotamia, however, was a system of contending states with much in common linguistically and culturally but no central control. These are the circumstances in which creativity flourishes best: Compare the Greek city-states of the Golden Age, or Renaissance Italy, or 19th-century Europe. Unification was occasional and short lived. No doubt the times were "interesting." Perhaps that is the price of creativity.

§1.2   One of the more impressive of these episodes of imperial unification in Mesopotamia ran from about 1790 to 1600 BCE. The unifier was Hammurabi, who came to power in the city-state of Babylon, on the middle Euphrates, around the earlier of those dates. Hammurabi[10] was an Amorite, speaking a dialect of Akkadian. He brought all of Mesopotamia under his rule and made Babylon the great city of the age. This was the first Babylonian empire.[11]

This first Babylonian empire was a great record-keeping civilization. Their writing was in the style called cuneiform, or wedge-shaped. That is to say, written words were patterns made by pressing a wedge-shaped stylus into wet clay. These impressed clay tablets and cylinders were baked for permanent record-keeping. Cuneiform had been invented by the Sumerians long before and adapted to Akkadian in the age of Sargon. By Hammurabi's time this writing method had evolved into a system of more than 600 signs, each representing an Akkadian syllable.

Here is a phrase in Akkadian cuneiform, from the preamble to Hammurabi's Code, the great system of laws that Hammurabi imposed on his empire.

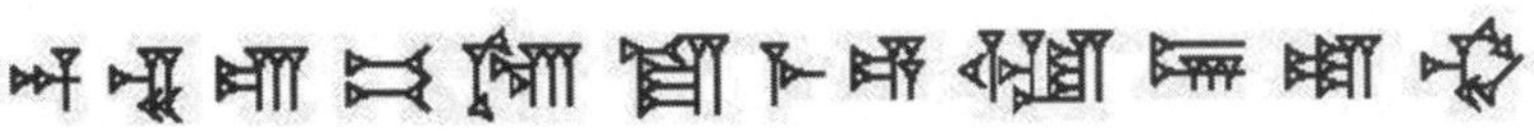

FIGURE 1-2 Cuneiform writing.

It would be pronounced something like *En-lil be-el sa-me-e u er-sce-tim*, meaning "Enlil, lord of heaven and earth." The fact that this is a Semitic language can be glimpsed from the word *be-el*, related to the beginning of the English "Beelzebub," which came to us from the Hebrew *Ba'al Zebhubh*—"Lord of the flies."

Cuneiform writing continued long after the first Babylonian empire had passed away—down to the 2nd century BCE, in fact. It was used for many languages of the ancient world. There are cuneiform inscriptions on some ruins in Iran, belonging to the dynasty of Cyrus the Great, around 500 BCE. These inscriptions were noticed by modern European travelers as long ago as the 15th century. Beginning in the late 18th century, European scholars began the attempt to decipher these inscriptions.[12] By the 1840s a good base of understanding of cuneiform inscriptions had been built up.

At about that same time, archeologists such as the Frenchman Paul Émile Botta and the Englishman Sir Austen Henry Layard were beginning to excavate ancient sites in Mesopotamia. Among the discoveries were great numbers of baked clay tablets inscribed with cuneiform. This archeological work has continued to the present day, and we now have over half a million of these tablets in public and private collections around the world, their dates ranging from the very beginning of writing around 3350 BCE to the 1st century BCE.

There is a large concentration of excavated tablets from the Hammurabi period, though, and for this reason the adjective "Babylonian" gets loosely applied to anything in cuneiform, although the first Babylonian empire occupied less than 2 of the 30-odd centuries that cuneiform was in use.

§1.3 It was known from early on—at least from the 1860s—that some of the cuneiform tablets contained numerical information. The first such items deciphered were what one would expect from a well-organized bureaucracy with a vigorous mercantile tradition: inventories, accounts, and the like. There was also a great deal of calendrical material. The Babylonians had a sophisticated calendar and an extensive knowledge of astronomy.

By the early 20th century, though, there were many tablets whose content was clearly mathematical but which were concerned with neither timekeeping nor accounting. These went mainly unstudied until 1929, when Otto Neugebauer turned his attention to them.

Neugebauer was an Austrian, born in 1899. After serving in World War I (which he ended in an Italian prisoner-of-war camp alongside fellow-countryman Ludwig Wittgenstein), he first became a physicist, then switched to mathematics, and studied at Göttingen under Richard Courant, Edmund Landau, and Emmy Noether—some of the biggest names in early 20th-century math. In the mid-1920s, Neugebauer's interest turned to the mathematics of the ancient world. He made a study of ancient Egyptian and published a paper about the Rhind Papyrus, of which I shall say more in a moment. Then he switched to the Babylonians, learned Akkadian, and embarked on a study of tablets from the Hammurabi era. The fruit of this work was the huge three-volume *Mathematische Keilschrift-Texte* (the German word *keilschrift* means "cuneiform") of 1935–1937, in which for the first time the great wealth of Babylonian mathematics was presented.

Neugebauer left Germany when the Nazis came to power. Though not Jewish, he was a political liberal. Following the purging of Jews from the Mathematical Institute at Göttingen, Neugebauer was appointed head of the institute. "He held the famous chair for exactly one day, refusing in a stormy session in the Rector's office to sign the required loyalty declaration," reports Constance Reid in her book *Hilbert*. Neugebauer first went to Denmark and then to the United States, where he had access to new collections of cuneiform tablets. Jointly with the American Assyriologist Abraham Sachs, he published *Mathematical Cuneiform Texts* in 1945, and this has remained a standard English-language work on Babylonian mathematics. Investigations have of course continued, and the brilliance of the Babylonians is now clear to everyone. In particular, we now know that they were masters of some techniques that can reasonably be called algebraic.

§1.4   Neugebauer discovered that the Hammurabi-era mathematical texts were of two kinds: "table texts" and "problem texts." The table texts were just that—lists of multiplication tables, tables of squares and cubes, and some more advanced lists, like the famous Plimpton 322 tablet, now at Yale University, which lists Pythagorean triples (that is, triplets of numbers a, b, c, satisfying $a^2 + b^2 = c^2$, as the sides of a right-angled triangle do, according to Pythagoras's theorem).

The Babylonians were in dire need of tables like this, as their system for writing numbers, while advanced for its time, did not lend itself to arithmetic as easily as our familiar 10 digits. It was based on 60 rather than 10. Just as our number "37" denotes three tens plus seven ones, the Babylonian number "37" would denote three sixties and seven ones—in other words, our number 187. The whole thing was made more difficult by the lack of any zero, even just a "positional" one—the one that, in our system, allows us to distinguish between 284, 2804, 208004, and so on.

Fractions were written like our hours, minutes, and seconds, which are ultimately of Babylonian origin. The number two and a half, for example, would be written in a style equivalent to "2:30." The Babylonians knew that the square root of 2 was, in their system, about 1:24:51:10. That would be $1 + (24 + (51 + 10 \div 60) \div 60) \div 60$, which is accurate to 6 parts in 10 million. As with whole numbers, though, the lack of a positional zero introduced ambiguities.

Even in the table texts, an algebraic cast of mind is evident. We know, for example, that the tables of squares were used to aid multiplication. The formula

$$ab = \frac{(a+b)^2 - (a-b)^2}{4}$$

reduces a multiplication to a subtraction (and a trivial division). The Babylonians knew this formula—or "knew" it, since they had no way to express abstract formulas in that way. They knew it as a procedure—we would nowadays say an *algorithm*—that could be applied to specific numbers.

§1.5   These table texts are interesting enough in themselves, but it is in the problem texts that we see the real beginnings of algebra. They contain, for example, solutions for quadratic equations and even for certain cubic equations. None of this, of course, is written in anything resembling modern algebraic notation. Everything is done with word problems involving actual numbers.

To give you the full flavor of Babylonian math, I will present one of the problems from *Mathematical Cuneiform Texts* in three formats: the actual cuneiform, a literal translation, and a modern working of the problem.

The actual cuneiform is presented in Figure 1-3. It is written on the two sides of a tablet, which I am showing here beside one another.[13]

Neugebauer and Sachs translate the tablet as follows: Italics are Akkadian; plain text is Sumerian; bracketed parts are unclear or "understood."

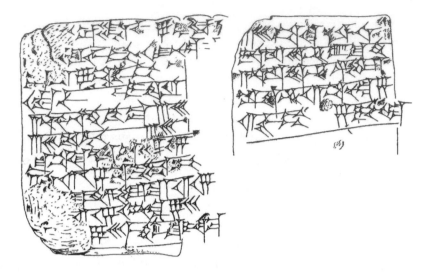

FIGURE 1-3 A problem text in cuneiform.

(Left of picture)
[*The* igib]um *exceeded the* igum *by* 7.
What are [*the* igum *and*] *the* igibum?
As for you—halve 7, *by which the* igibum *exceeded the* igum,
     and (the result is) 3;30.
Multiply together 3;30 *with* 3;30, *and (the result is)* 12;15.
To 12;15, *which resulted for you,*
add [1,0, *the produ*]*ct, and (the result is)* 1,12;15.
What is [*the square root of* 1],12;15? (*Answer:*) 8;30.
Lay down [8;30 *and*] 8;30, *its equal, and then*

(Right of picture)
Subtract 3;30, *the item, from the one,*
add (it) to the other.
One is 12, *the other* 5.
12 *is the* igibum, 5 *the* igum.

(Note: Neugebauer and Sachs are using commas to separate the "digits" of numbers here, with a semicolon to mark off the whole number part from the fractional part of a number. So "1,12;15" means $1 \times 60 + 12 + \frac{15}{60}$, which is to say, $72\frac{1}{4}$.)

Here is the problem worked through in a modern style:

A number exceeds its reciprocal by 7. Note, however, that because of the place-value ambiguity in Babylonian numerals, the "reciprocal" of x may mean $\frac{1}{x}$, or $\frac{60}{x}$, or $\frac{3600}{x}$ ... in fact, any power of 60 divided by x. It seems from the solution that the authors have taken "reciprocal" here to mean $\frac{60}{x}$. So

$$ x - \frac{60}{x} = 7 $$

What are x and its "reciprocal"? Since the equation simplifies to

$$ x^2 - 7x - 60 = 0 $$

we can apply the familiar formula[14] to get

$$ x = \frac{7 \pm \sqrt{7^2 + (4 \times 60)}}{2} $$

This delivers solutions $x = 12$ and $x = -5$. The Babylonians knew nothing of negative numbers, which did not come into common use until 3,000 years later. So far as they were concerned, the only solution is 12; and its "reciprocal" (that is, $\frac{60}{x}$) is 5. In fact, their algorithm does not deliver the two solutions to the quadratic equation, but is equivalent to the slightly different formula

$$ x = \sqrt{\left(\frac{7}{2}\right)^2 + 60} \pm \frac{7}{2} $$

for x and its "reciprocal." You might, if you wanted to be nitpicky about it, say that this means they did not, strictly speaking, solve the quadratic equation. You would still have to admit, though, that this is a pretty impressive piece of early Bronze Age math.

§1.6   I emphasize again that the Babylonians of Hammurabi's time had no proper algebraic symbolism. These were word problems, the quantities expressed using a primitive numbering system. They had taken only a step or two toward thinking in terms of an "unknown quantity," using Sumerian words for this purpose in their Akkadian text, like the *igum* and *igibum* in the problem above. (Neugebauer and Sachs translate both *igum* and *igibum* as "reciprocal." In other contexts the tablets use Sumerian words meaning "length" and "width," that is, of a rectangle.) The algorithms supplied were not of universal utility; different algorithms were used for different word problems.

Two questions arise from all this. First: Why did they bother? Second: Who first worked this all out?

Regarding the first question, the Babylonians did not think to tell us why they were doing what they were doing. Our best guess is that these word problems arose as a way to check calculations—calculations involving measurement of land areas or questions involving the amount of earth to be moved to make a ditch of certain dimensions. Once a rectangular field had been marked out and its area computed, you could run area and perimeter "backward" through one of these quadratic equation algorithms to make sure you got the numbers right.

To the second, the proto-algebra in the Hammurabi-era tablets is mature. From what we know of the speed of intellectual progress in remote antiquity, these techniques must have been cooking for centuries. Who first thought them up? This we do not know, though the use of Sumerian in these problem tablets suggests a Sumerian origin. (Compare the use of Greek letters in modern mathematics.) We have texts going back before the Hammurabi era, deep into the third millennium, but they are all arithmetical. Only at this time, the 18th and 17th centuries BCE, does algebraic thinking show up. If there were "missing link" texts that show an earlier development of these algebraic ideas, they have not survived, or have not yet been found.

Nor do the Hammurabi-era tablets tell us anything about the people who wrote them. We know a great deal about Babylonian math, but we don't know any Babylonian mathematicians. The first person whose name we know, and who was very likely a mathematician, lived at the other end of the Fertile Crescent.

§1.7   While the Hammurabi dynasty was consolidating its rule over Mesopotamia, Egypt was enduring its first foreign invasion. The invaders were a people known to us by the Greek word *Hyksos*, a corruption of an Egyptian phrase meaning "rulers of foreign lands." Moving in from Palestine, not in a sudden rush, but by creeping annexation and colonization, they had established a capital at Avaris, in the Eastern Nile delta, by around 1720 BCE.

During the Hyksos dynasty there lived a man named Ahmes, who has the distinction of being the first person whose name we know and who has some definite connection with mathematics. Whether Ahmes was actually a working mathematician is uncertain. We know of him from a single papyrus, dating from around 1650 BCE—the early part of the Hyksos dynasty. In that papyrus, Ahmes tells us he is acting as a scribe, copying a document written in the Twelfth Dynasty (about 1990–1780 BCE). Perhaps this was one of the text preservation projects that we know were initiated by the Hyksos rulers, who were respectful of the then-ancient Egyptian civilization. Perhaps Ahmes was a mathematical ignoramus, blindly copying what he saw. This, however, is unlikely. There are few mathematical errors in the papyrus, and those that exist look much more like errors in computation (wrong numbers being carried forward) than errors in copying.

This document used to be called the Rhind Papyrus, after A. Henry Rhind, a Scotsman who was vacationing in Egypt for his health—he had tuberculosis—in the winter of 1858. Rhind bought the papyrus in the city of Luxor; the British Museum acquired it when he died five years later. Nowadays it is thought more proper to name the papyrus after the man who wrote it, rather than the man who bought it, so it is now usually called the Ahmes Papyrus.

While mathematically fascinating and a great find, the Ahmes Papyrus contains only the barest hints of algebraic thinking, in the sense I am discussing. Here is Problem 24, which is as algebraic as the papyrus gets: "A quantity added to a quarter of itself makes 15." We write this in modern notation as

$$x + \frac{1}{4}x = 15$$

and solve for the unknown x. Ahmes adopted a trial-and-error approach—there is little in the way of Babylonian-style systematic algorithms in the papyrus.

§1.8   "A considerable difference of opinion exists among students of ancient science as to the caliber of Egyptian mathematics," wrote James R. Newman in *The World of Mathematics*. The difference apparently remains. After looking over representative texts from Babylonia and Egypt, though, I don't see how anyone could maintain that these two civilizations, flourishing at opposite ends of the Fertile Crescent in the second quarter of the second millennium BCE, were equal in mathematics. Though both were working in arithmetical styles, with little evidence of any powers of abstraction, the Babylonian problems are deeper and more subtle than the Egyptian ones. (This was also Neugebauer's opinion, by the way.)

It is still a wonderful thing that with only the most primitive methods for writing numbers, these ancient peoples advanced as far as they did. Perhaps even more astonishing is the fact that they advanced very little further in the centuries that followed.

Chapter 2

THE FATHER OF ALGEBRA

§2.1 FROM EGYPT TO EGYPT: Diophantus,[15] the father of algebra, in whose honor I have named this chapter, lived in Alexandria, in Roman Egypt, in either the 1st, the 2nd, or the 3rd century CE.

Whether Diophantus actually *was* the father of algebra is what lawyers call "a nice point." Several very respectable historians of mathematics deny it. Kurt Vogel, for example, writing in the *DSB*, regards Diophantus's work as not much more algebraic than that of the old Babylonians and Archimedes (3rd century BCE; see §2.3 below), and concludes that "Diophantus certainly was not, as he has often been called, the father of algebra." Van der Waerden pushes the parentage of algebra to a point *later* in time, beginning with the mathematician al-Khwarizmi, who lived 600 years after Diophantus and whom I shall get to in the next chapter. Furthermore, the branch of mathematics known as Diophantine analysis is most often taught to modern undergraduates as part of a course in number theory, not algebra.

I shall give an account of Diophantus's work and let you make your own judgment, offering my opinion on the matter, for what it is worth, as a conclusion.

§2.2   The people of Mesopotamia went on writing in cuneiform through centuries of ethnic and political churning, down to the region's conquest by the Parthians in 141 BCE. We have mathematical texts in cuneiform from right up to that conquest. It is an astonishing thing, testified to by everyone who has studied this subject, that there was almost no progress in mathematical symbolism, technique, or understanding during the millennium and a half that separates Hammurabi's empire from the Parthian conquest. The mathematician John Conway, who has made a study of cuneiform tablets, says that the only difference that presents itself to the eye is a "positional zero" marker in the latest tablets—a way, that is, to distinguish between, say, 281 and 2801. As with Mesopotamia, so with Egypt: We have no grounds for thinking that Egyptian mathematics made any notable progress from the 16th to the 4th century BCE.

If the mathematicians of Babylon and Egypt had made no progress in their homelands, though, their brilliant early discoveries had spread throughout the ancient West, and possibly beyond. From this point on—in fact, from the 6th century BCE—the story of algebra in the ancient world is a Greek story.

§2.3   The peculiarity of Greek mathematics is that prior to Diophantus it was mainly geometrical. The usual reason given for this, which sounds plausible to me, is that the school of Pythagoras (late 6th century BCE) had the idea to found all mathematics—and music, and astronomy—on number but that the discovery of irrational numbers so disturbed the Pythagoreans, they turned away from arithmetic, which seemed to contain numbers that could not be written, to geometry, where such numbers could be represented infallibly by the lengths of line segments.

Early Greek algebraic notions were therefore expressed in geometrical form, often very obscurely. Bashmakova and Smirnova, for example, note that Propositions 28 and 29 in Book 6 of Euclid's great treatise *The Elements* offer solutions to quadratic equations. I suppose they do, but this is not, to say the least of it, obvious on a first reading. Here is Euclid's Proposition 28, in Sir Thomas Heath's translation:

> To a given straight line to apply a parallelogram equal to a given rectilineal figure and deficient by a parallelogrammic figure similar to a given one: thus the given rectilineal figure must not be greater than the parallelogram described on the half of the straight line and similar to the defect.

Got that? This, say Bashmakova and Smirnova, is tantamount to solving the quadratic equation $x(a - x) = S$. I am willing to take their word for it.

Euclid lived, taught, and founded a school in the city of Alexandria when it, and the rest of Egypt, was ruled by Alexander's general Ptolemy (whose regnal dates as Ptolemy I are 306–283 BCE). Alexandria had been founded—by Alexander, of course—on the western edge of the Nile delta, looking across the Mediterranean to Greece, shortly before Euclid was born. Euclid himself is thought to have gotten his mathematical training in Athens, at the school of Plato, before settling in Egypt. Be that as it may, Alexandria in the 3rd century BCE was a great center of mathematical excellence, more important than Greece itself.

Archimedes, who was 40 years younger than Euclid, and who probably studied under Euclid's successors in Alexandria, continued the geometric approach, though taking it into much more difficult terrain. His book *On Conoids and Spheroids*, for example, treats the intersection of a plane with curved two-dimensional surfaces of a sophisticated kind. It is clear from this work that Archimedes could solve certain particular kinds of cubic equations, just as Euclid could solve some quadratic equations, but all the language is geometric.

§2.4 Alexandrian mathematics went into something of a decline after the glory days of the 3rd century BCE, and in the disorderly 1st century BCE (think of Anthony and Cleopatra) seems to have died out altogether. With the more settled conditions of the early Roman imperial era, there was a revival. There was also a turn of thought away from purely geometrical thinking, and it was in this new era that Diophantus lived and worked.

As I intimated at the beginning of this chapter, we know next to nothing about Diophantus, not even the century in which he lived.

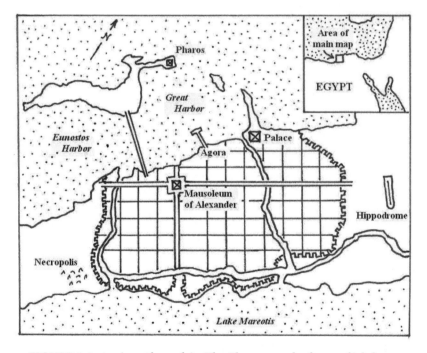

FIGURE 2-1 Ancient Alexandria. The Pharos was the famous lighthouse, one of the Seven Wonders of the ancient world, destroyed by a succession of earthquakes from the 7th century to the 14th. The great library is thought to have been in the vicinity of the palace, in the northeastern part of the city. The crenellated line shows the original (331 BCE) city walls.

The 3rd century CE is the most popular guess, with the dates 200–284 CE often quoted. Diophantus's claim on our attention is a treatise he wrote, titled *Arithmetica*, of which less than half has come down to us. The main surviving part of the treatise consists of 189 problems in which the object is to find numbers, or families of numbers, satisfying certain conditions. At the beginning of the treatise is an introduction in which Diophantus gives an outline of his symbolism and methods.

The symbolism looks primitive to our eyes but was very sophisticated in its own time. An example will show the main points. Here is an equation in modern form:

$$x^3 - 2x^2 + 10x - 1 = 5$$

Here it is as Diophantus writes it:

$$K^Y \bar{\alpha} \varsigma \bar{\imath} \ \pitchfork \ \Delta^Y \bar{\beta} \ \mathbf{M} \ \bar{\alpha} \ \acute{\imath}\sigma \ \mathbf{M} \bar{\varepsilon}$$

The easiest thing to pick out here is the numbers. Diophantus used the Greek "alphabetic" system for writing numbers. This worked by taking the ordinary Greek alphabet of 24 letters and augmenting it with three obsolete letters to give a total of 27 symbols. These were divided into three groups of nine each. The first nine letters of this augmented alphabet represented the digits from 1 to 9; the second nine represented the tens from 10 to 90; the third nine represented the hundreds from 100 to 900. The Greeks had no symbol for zero; neither did anyone else at this time.[16]

So in the equation I have shown, $\bar{\alpha}$ represents 1, $\bar{\beta}$ represents 2, $\bar{\imath}$ represents 10, and $\bar{\varepsilon}$ represents 5. (The bars on the top are just to show that these letters are being used as numerals.)

Of the other symbols, $\acute{\imath}\sigma$ is short for $\acute{\imath}\sigma o\varsigma$, meaning "equals." Note that there are no bars on top here; these are letters being used to spell a word (actually an abbreviation), not to represent numbers. The inverted trident, $\pitchfork$, indicates subtraction of everything that follows it, as far as the "equals sign."

That leaves us with four symbols left to explain: K^Y, ς, Δ^Y, and $\dot{M}$. The second one, ς, is the unknown quantity, our modern x. The others are powers of the unknown quantity: K^Y the third power (from Greek $\kappa\acute{v}\beta o\varsigma$, a cube), Δ^Y the square (from $\delta\acute{v}\nu\alpha\mu\iota\varsigma$, strength or power), $\dot{M}$ the zeroth power—what we would nowadays call "the constant term."

Armed with this knowledge, we can make a literal, symbol-for-symbol, translation of Diophantus's equation as follows:

$$x^3 1 \; x10 - x^2 2 \; x^0 1 = x^0 5$$

It makes a bit more sense if you put in the understood plus signs and some parentheses:

$$(x^3 1 + x10) - (x^2 2 + x^0 1) = x^0 5$$

Since Diophantus wrote the coefficient after the variable, instead of before it as we do (that is, instead of "$10x$," he wrote "$x10$"), and since any number raised to the power of zero is 1, this is equivalent to the equation as I originally wrote it:

$$x^3 - 2x^2 + 10x - 1 = 5$$

It can be seen from this example that Diophantus had a fairly sophisticated algebraic notation at his disposal. It is not clear how much of this notation was original with him. The use of special symbols for the square and cube of the unknown was probably Diophantus's own invention. The use of ς for the unknown quantity, however, seems to have been copied from an earlier writer, author of the so-called Michigan Papyrus 620, in the collection of the University of Michigan.[17]

Diophantus's system of notation had some drawbacks, too. The principal one was that he could not represent more than one unknown. To put it in modern terms, he had an x but no y or z. This is a major difficulty for Diophantus, because most of his book (though not, as Gauss wrongly said, all of it), deals with *indeterminate equations*. This needs a little explanation.

§2.5   The word "equation" as mathematicians use it just means the statement that something is equal to something else. If I say "two plus two equals four," I have stated an equation. Of course, the equations that mathematicians, including Diophantus, are interested in are ones with some unknown quantities in them. The presence of an unknown moves the equation from the indicative mood, "this is so," to the interrogative, "is this so?" or, more often, "*when* is this so?" The equation

$$x + 2 = 4$$

is implicitly a question: "*What* plus two equals four?" The answer is of course 2. This equation is so when $x = 2$.

Suppose, however, I ask this question:

$$x + y + 2 = 4$$

What is the answer? We are now in deeper waters.

In the first place, a mathematician will immediately want to know what *kind* of answers you are seeking. Positive whole numbers only? Then the only solution is $x = 1, y = 1$. Will you settle for non-negative whole numbers (that is, including zero)? Now we have two more solutions: (a) $x = 0, y = 2$, and (b) $x = 2, y = 0$. Will you allow negative numbers as solutions? If you will, there is now an *infinity* of solutions—this one, for example: $x = 999, y = -997$. Will you allow rational numbers? If so, again there is an infinity of solutions, like this one: $x = \frac{157}{111}, y = \frac{65}{111}$. And, of course, if you permit irrational numbers or complex numbers as solutions, then infinities pile upon infinities.

Equations of this sort, with more than one unknown in them and a potentially infinite number of solutions (depending on what *kind* of solutions are asked for), are called *indeterminate*.

The most famous of all indeterminate equations is the one that occurs in Fermat's Last Theorem:

$$x^n + y^n = z^n$$

where x, y, z, and n must all be positive whole numbers. This equation has an infinity of solutions if $n = 1$ or $n = 2$. Fermat's Last Theorem states that it has no solutions when n is greater than 2.

Pierre de Fermat was reading Diophantus's *Arithmetica* (in a Latin translation), around the year 1637, when the theorem occurred to him, and it was in the margin of this book that he made his famous note, stating the theorem, then adding (also in Latin): "Of which I have discovered a really wonderful proof, but the smallness of the margin cannot contain it." The theorem was actually proved 357 years later, by Andrew Wiles.

§2.6   As I said, most of the *Arithmetica* consists of indeterminate equations. And, as I also said, this put Diophantus at a grave disadvantage, since he had a symbol for only one unknown (with those extra symbols for its square, cube, and so on).

To see how he got around this difficulty, here is his solution to the problem alongside which Fermat wrote that famous marginal note: Problem 8 in Book 2.

Diophantus states the problem: "A square number is to be decomposed into a sum of squares." We should nowadays express this problem as: "Given a number a, find numbers x and y such that $x^2 + y^2 = a^2$." Diophantus, of course, did not have a notation as sophisticated as ours, so he preferred to spell out the problem in words.

To solve the problem, he begins by giving a a definite value, $a = 4$. So we are seeking x and y for which $x^2 + y^2 = 16$. He then writes y as an expression in x, the particular, though apparently arbitrary, expression $y = 2x - 4$. So now we have a specific equation to solve, one to which Diophantus could apply his literal symbolism:

$$x^2 + (2x - 4)^2 = 16$$

This is just a quadratic equation, and Diophantus knew how to solve it. The solution is $x = \frac{16}{5}$. (There is another solution: $x = 0$. Diophantus, who had no symbol for zero, ignores this.) It follows that $y = \frac{12}{5}$.

This does not look very impressive. In fact, it looks like cheating. The equation $x^2 + y^2 = a^2$ has an infinity of solutions. Diophantus only got one. He got it by a procedure that can easily be generalized, though, and he knew—he says it elsewhere in his book—that there are infinitely many other solutions.

§2.7   I noted before that the first question in a mathematician's mind when faced with an indeterminate equation like $x + y + 2 = 4$ is: What *kind* of solutions are you looking for? In Diophantus's case the answer is positive rational numbers, like the $\frac{16}{5}$ and $\frac{12}{5}$ of the problem I worked out a moment ago. Negative numbers, along with zero, had not yet been discovered; Diophantus remarks of the equation $4x + 20 = 4$ that it is "absurd." He knows about irrational numbers, of course, but displays no interest in them. When they show up in a problem, he adjusts the terms of the problem to get rational solutions.

Seeking rational number solutions to problems like the ones Diophantus tackled is really equivalent to seeking whole-number solutions to very closely related problems. The equality

$$\left(\frac{16}{5}\right)^2 + \left(\frac{12}{5}\right)^2 = 4^2$$

is really

$$16^2 + 12^2 = 20^2$$

in thin disguise. So nowadays we use the phrase "Diophantine analysis" to mean "seeking whole-number solutions to polynomial equations."

§2.8   Possibly the reader is still unimpressed by Diophantus's attack on the equation $x^2 + y^2 = a^2$. There seems to be remarkably little there to show for 2,000 years of mathematical reflection on the Babylonians' solution of the quadratic equation and their other achievements.

In fairness to Diophantus, I should say that while that particular problem is a handy one to bring forward, he tackled much harder problems than that. The equation $x^2 + y^2 = a^2$ illustrates his methods very well, it has that interesting connection with Fermat's Last Theorem, and it doesn't take long to explain; that is why it's a popular example. Elsewhere Diophantus takes on cubic and quartic equations in a single unknown; systems of simultaneous equations in two, three, and four unknowns; and a problem equivalent to a system of eight equations in 12 unknowns.

Diophantus knew much more than is on display in his solution of $x^2 + y^2 = a^2$. He knew the rule of signs, for example, stating it thus:

> Wanting (that is, negativity) multiplied by wanting yields forthcoming (that is, positivity); wanting multiplied by forthcoming equals wanting.

This is a pretty remarkable thing, considering that, as I have already said, negative numbers had not yet been discovered!

In fact, that "not discovered" needs some slight qualification. While Diophantus had no notion of, let alone any notation for, negative numbers as independent mathematical objects and would not admit them as solutions, he actually makes free use of them *inside* his computations, for example, subtracting $2x + 7$ from $x^2 + 4x + 1$ to get $x^2 + 2x - 6$. Clearly, even though he regarded –6 to be "absurd" as a mathematical object in its own right, he nonetheless knew, at some level, that $1 - 7 = -6$.

It is situations like this that make us realize how deeply *unnatural* mathematical thinking is. Even such a basic concept as negative numbers took centuries to clarify itself in the minds of mathematicians, with many intermediate stages of awareness like this. Something similar happened 1,300 years later with imaginary numbers.

Diophantus also knew how to "bring a term over" from one side of an equation to the other, changing its sign as he did so; how to "gather up like terms" for simplification; and some elementary principles of expansion and factorization.

His insistence on rational solutions also pointed into the far future, to the algebraic number theory of our own time. That equation $x^2 + y^2 = a^2$ is, we would say nowadays, the equation of a circle with radius a. In seeking rational-number solutions to it, Diophantus is asking: Where are the points on that circle having rational-number coordinates x and y? That is a very modern question indeed, as we shall see much later (§14.4).

§2.9   So, was Diophantus the father of algebra? I am willing to give him the title just for his literal symbolism—his use of special letter symbols for the unknown and its powers, for subtraction and equality. The first time I saw one of Diophantus's equations laid out in his own symbolism, my reaction was, as yours probably was: "Say *what*?" Looking through some of his problems, though, I quickly got used to it, to the degree that I could "read off" a Diophantine equation without much pause for thought.

At last I appreciated what a great advance Diophantus made with his literal symbolism. I do take Vogel's point about the absence of general methods in the *Arithmetica* and am willing to take on faith Diophantus's lack of originality in choice of topics. Probably it is also the case that he was not the first to deploy a special symbol for the unknown quantity.

Diophantus was, however, by the fortunes of history, the earliest to pass down to us such a widely comprehensive collection of problems so imaginatively presented. It is a shame we do not know who first used a symbol for the unknown, but since Diophantus used it so well, so early, we ought to honor him for that. Probably someone of whom we have no knowledge, nor ever will have any knowledge, was the true father of algebra. Since the title is vacant, though, we may as well attach it to the most worthy name that has survived from antiquity, and that name is surely Diophantus.

FIGURE 2-2 The Pharos (lighthouse) at Alexandria, as imagined by
Martin Heemskerck (1498–1574).

Chapter 3

COMPLETION AND REDUCTION

§3.1 THE WORD "ALGEBRA," as everyone knows, comes from the Arabic language. This has always seemed to me somewhat unfair, for reasons I shall get to shortly. Unfair or not, it requires some explanation from the historian.

§3.2 Supposing the dates I gave for Diophantus (200–284 CE) are correct, he lived through a very unhappy period. The Roman Empire, of which Egypt was a province, was then embarked on its well-known decline and fall, of which Edward Gibbon wrote so eloquently and at such length. The sorry state of affairs in Diophantus's time, if it was his time, is described in Chapter 7 of Gibbon's masterpiece.

The empire rallied somewhat in the later 3rd century. Diocletian (284–305 CE) and Constantine (306–337 CE) are counted among the great Roman emperors. The first of them fiercely persecuted Christians; the second was the son of a Christian, issued the Edict of Milan (313 CE) commanding tolerance of Christianity throughout the empire, and himself accepted baptism in his last illness.

The decision first to tolerate and then to enforce the practice of Christianity did little to retard the crumbling of the empire. In some ways it may have accelerated it. One of the great strengths of early Christianity was that it appealed to all classes. To do this, however, it had to "pay off" the sophisticated urban intellectuals with appealing but complex metaphysical theories, while maintaining its hold on the masses with a plain, clear message of salvation and divine powers, fortified with some colorful stories and concessions to older pagan beliefs. Inevitably, though, the masses got wind of the lofty metaphysical disputes and used them as fronts for social and ethnic grievances.

Diophantus's own city of Alexandria illustrates the process well. Even after 300 years as a Greek city and a further 300 as a Roman one, Alexandria remained a glittering urban enclave that was fed and clothed from a hinterland of illiterate, Coptic-speaking Egyptian peasants. To a Christian Copt from the desert fringes, the words "Greek," "Roman," and "Pagan" must have been near-synonyms, and the fabulous Museion ("Temple of the Muses"), with its tradition of secular learning and its attached great library, probably seemed to be a house of Satan.

The matter was made worse in Egypt by the cult of monasticism, especially strong there, which placed several thousand vigorous but sex-starved young males at the service of anyone who wanted to whip up a religious mob. Which, of course, ambitious politicians—a category that at this point in Roman history frequently included officers of the church—often did. This is the context for the murder of Hypatia in 415 CE, which so outraged Gibbon.

Hypatia is the first female name in the history of mathematics. All her written works have been lost, so we know them only by hearsay. On this basis it is difficult to judge whether she can properly be called a significant mathematician or not. It is certain, at any rate, that she was a major intellectual. She taught at the Museion (of which her father, Theon, was the last president) and was a compiler, editor, and preserver of texts, including math texts. She was an adherent and a teacher of the philosophy called Neoplatonism, an attempt to locate in another world the order, justice, and peace so conspicuously lacking in the later Roman empire.[18] She was also, we are told, a beauty and a virgin.

Hypatia was active in scholarly teaching and learning during the time when Cyril of Alexandria was archbishop of the city. Cyril, later Saint Cyril, is a difficult person to judge through the mists of time and theological controversy. Gibbon gives him very unfavorable coverage, but Gibbon nursed a prejudice against Christianity and can't be altogether trusted. Certainly Cyril launched a pogrom against Alexandria's Jews, driving them out of the city, but the Jews seem previously to have conducted a nasty anti-Christian pogrom of their own, as even Gibbon allows.[19] The Alexandrians of this time were, as we learn from the *Catholic Encyclopedia*, "always riotous." At any rate, Cyril got into a church-state dispute with Orestes, the Prefect (that is, the Roman official in charge of Egypt), and it was put about that Hypatia was the main obstacle to the healing of this split. A mob was raised, or raised itself, and Hypatia was pulled from her chariot and dragged through the streets to a church, where the flesh was scraped from her bones with, depending on the authority and translator, either oyster shells or pottery shards.[20]

Hypatia seems to have been the last person to teach at the Museion, and her appalling death in 415 is usually taken to mark the end of mathematics in the ancient European world. The Roman empire in the west limped along for another 60 years, and Alexandria continued under the authority of the eastern, Byzantine, emperors for a further 164 years (interrupted by a brief Persian[21] occupation, 616–629), but the intellectual vitality was all gone. The next noteworthy name in the history of algebra had his home 900 miles due east of Alexandria, on the banks of the Tigris, in that same Mesopotamian plain where the whole thing had begun two and a half millennia before.

§3.3   The northern and western territories of the Roman empire were lost to Germanic barbarians in the 5th century. The southern and eastern ones, except for Greece, Anatolia, and some southern fragments of Italy and the Balkans, fell to the armies of Islam in the 7th. Alexandria itself fell on December 23, 640 CE, breaking the heart of the Byzantine emperor Heraclius, who had spent his entire working life recovering territories that had been lost to the early 7th-century Persian resurgence.[22]

The actual conqueror of Alexandria was a person named Amr ibn al-'As.[23] He was a direct report to Omar, the second Caliph (that is, leader of the Muslims after Mohammed's death). Alexandria put up a fight—the siege lasted 14 months—but most of Egypt was a pushover for the Muslims. The Egyptians had persisted in a Christian heresy called Monophysitism, and after wresting the province back from its brief subjection to Persia, the Byzantine emperor Heraclius persecuted them savagely for this.[24] The result was that the native Egyptians came to detest the Byzantines and were glad to exchange a harsh master for a more tolerant one.

The third Caliph, Othman, belonged to the Omayyad branch of Mohammed's clan. After some complicated civil wars between his faction and that of the Prophet's son-in-law Ali (one indirect result of which was the Sunni–Shiite split that still divides Islam today), the Omayyads established a dynasty that ruled the Islamic world from Damascus for 90 years, 661–750 CE. Then a revolt led to a change of dynasty, the Omayyads keeping only Spain, where they hung on for another 300 years.

The new dynasty traced its descent to Mohammed's uncle al-Abbas, and so its rulers are known to history as the Abbasids. They founded a new capital, Baghdad, in 762 CE, plundering the old Babylonian and Persian ruins for building materials. The English word "algebra" is taken from the title of a book written in this Baghdad of the Abbasid dynasty around 820 CE by a man who rejoiced in the name Abu Ja'far Muhammad ibn Musa al-Khwarizmi. I shall refer to him from here on just as al-Khwarizmi, as everyone else does.[25]

§3.4 Baghdad under the fifth, sixth, and seventh Abbasid Caliphs (that is, from 786 to 833 CE) was a great cultural center, dimly familiar to modern Westerners as the world of viziers, slaves, caravans, and far-traveling merchants pictured in the *Arabian Nights* stories. Arabs themselves look back on this as a golden age, though in fact the Caliphate no longer had the military strength to keep together all the conquests won in the first flush of Islamic vigor and was losing territories to rebels in North Africa and the Caucasus.

Persia was part of the Abbasid domain, under the spiritual and temporal authority of the Caliph. However, Persia had been the home of high civilizations since the Median empire of 1,400 years earlier, while the Arabs of 800 CE were only half a dozen generations away from their roots as desert-dwelling nobodies. The Abbasids therefore nursed something of a cultural inferiority complex toward the Persians, rather as the Romans did toward the Greeks.

Beyond the Persians were the Indians, whom the first great Muslim expansion left untouched. Northern India had been united under the Gupta dynasty in the 4th and 5th centuries CE but thereafter was generally divided into petty states until Turkish conquerors arrived in the late 10th century. These medieval Hindu civilizations were fascinated by numbers, especially very big numbers, for which they had special names. (The Sanskrit term *tallakchana*, should you ever encounter it, means a hundred thousand trillion trillion trillion trillion.) It is to the Indians—probably to the mathematician Brahmagupta, 598–670 CE—that the immortal honor of having discovered the number zero belongs, and our ordinary numerals, which we call Arabic, are actually of Indian origin.

Beyond the Indians were of course the Chinese, with whom India had been in cultural contact since at least the travels of the Buddhist monk Xuan-zang in the middle 7th century and with whom the Persians conducted a busy trade along the Silk Route. The Chinese had long had a mathematical culture of their own—I shall say something about it in §9.1.

Those inhabitants of Baghdad who had the leisure and inclination could therefore acquaint themselves with everything that was
known anywhere in the civilized world at that time. The culture of
the Greeks and Romans was familiar to them through Alexandria,
now one of their cities, and through their trading contacts with the
Byzantine empire. The cultures of Persia, India, and China were easily accessible.

All that was needed to make Abbasid Baghdad an ideal center for
the preservation and enrichment of knowledge was an academy, a
place where written documents could be consulted and lectures and
scholarly conferences held. Such an academy soon appeared. It was
called *Dar al-Hikma*, the "House of Wisdom." This academy's greatest flourishing was in the reign of the seventh Abbasid Caliph,
al-Mamun. In the words of Sir Henry Rawlinson, Baghdad under
al-Mamun "in literature, art, and science . . . divided the supremacy
of the world with Cordova; in commerce and wealth it far surpassed
that city." This was the time when al-Khwarizmi lived and worked.

§3.5   We know very little about al-Khwarizmi's life. His dates are
known only approximately. There are some fragmentary dry notices
in the works of Islamic historians and bibliographers, for details of
which I refer the reader to the *DSB*. We do know that he wrote several
books: one on astronomy, one on geography, one on the Jewish calendar, one on the Indian system of numerals, one a historical
chronicle.

The work on Indian numerals survives only in a later Latin translation, whose opening words are "*Dixit Algorithmi . . .*" ("According
to al-Khwarizmi . . ."). This book lays out the rules for computing
with the modern 10-digit place value system of arithmetic, which the
Indians had invented, and it was tremendously influential. Because of
those opening words, medieval European scholars who had mastered
this "new arithmetic" (as opposed to the old Roman numeral system,
which was hopeless for computation) called themselves
"algorithmists." Much later the word "algorithm" was used to mean

any process of computing in a finite number of well-defined steps. This is the sense in which modern mathematicians and computer scientists use it.

The book that really concerns us is the one titled *al-Kitab al-mukhtasar fi hisab al-jabr wa'l-muqabala* ("A Handbook of Calculation by Completion and Reduction"). This is a textbook of algebra and arithmetic, the first significant work in the field since Diophantus's *Arithmetica* of 600 years earlier. The book is in three parts, with these topics: solution of quadratic equations, measurement of areas and volumes, and the math required to deal with the very complicated Islamic laws of inheritance.

Only the first of these three parts is strictly algebraic, and it is something of a disappointment. For one thing, al-Khwarizmi has no literal symbolism—no way to lay out equations in letters and numbers, no sign for the unknown quantity and its powers. The equation we would write as

$$x^2 + 10x = 39$$

and which Diophantus would have written as

$$\Delta^Y \bar{\alpha} \; \varsigma \bar{\iota} \; '\iota\sigma \; \mathbf{M} \; \overline{\lambda\theta}$$

appears in al-Khwarizmi's book as

> One square and ten roots of the same amount to thirty-nine *dirhems*; that is to say, what must be the square which, when increased by ten of its own roots, amounts to thirty-nine?

(*Dirhem* was a unit of money. Al-Khwarizmi uses it to refer to what we nowadays call the constant term, the term in x^0.)

For another thing, Diophantus's historic turn away from the geometrical method toward manipulation of symbols is nowhere visible in al-Khwarizmi's work. This is not very surprising, since he had no symbols to manipulate, but it is still a sliding back from Diophantus's great breakthrough of 600 years earlier. Says van der Waerden: "[W]e may exclude the possibility that al-Khwarizmi's work was much influenced by classical Greek mathematics."

Al-Khwarizmi's main algebraic achievement, in fact, was to bring forward the idea of *equations* as objects of interest, classifying all equations of the first and second degrees in one unknown and giving rules for manipulating them. His classification was into six fundamental types, which we would write in modern symbolism as

(1) $ax^2 = bx$ (3) $ax = b$ (5) $ax^2 + c = bx$
(2) $ax^2 = b$ (4) $ax^2 + bx = c$ (6) $ax^2 = bx + c$

Some of these look trivially the same type to us but that is because we have negative numbers to help us. Al-Khwarizmi had no such aids. He could speak of subtraction, of course, and of one quantity exceeding another, or falling short of another, but his natural arithmetic tendency was to see everything in terms of positive quantities.

As for techniques of manipulation, that is where "completion" (*al-jabr*) and "reduction" (*al-muqabala*) come in. Once you have an equation like

$$x^2 = 40x - 4x^2$$

(or, as al-Khwarizmi says, "a square which is equal to forty things less four squares"), how do you manipulate it into one of those six standard forms? By *al-jabr*, that's how—"completing" the equation by adding $4x^2$ to each side, leaving us with a type-1 equation:

$$5x^2 = 40x.$$

That is adding equal terms on both sides. The opposite thing, where you need to *subtract* equal terms from both sides, is *al-muqabala*, for example, turning the equation

$$50 + x^2 = 29 + 10x$$

into a type 5

$$21 + x^2 = 10x,$$

by subtracting 29 from both sides.

§3.6   None of this is really new. In fact, *al-jabr* and *al-muqabala* can both be found in Diophantus—together with, of course, a rich literal symbolism to aid the manipulations. "Al-Khwarizmi's scientific achievements were at best mediocre, but they were uncommonly influential," says Toomer in the *DSB*.

I fear, in fact, that at this point the reader may be slipping into the conviction that these ancient and medieval algebraists were not very bright. We started in 1800 BCE with the Babylonians solving quadratic equations written as word problems, and now here we are 2,600 years later with al-Khwarizmi . . . solving quadratic equations written as word problems.

It is, I agree, all a bit depressing. Yet it is also inspiring, in a way. The extreme slowness of progress in putting together a symbolic algebra testifies to the very high level at which this subject dwells. The wonder, to borrow a trope from Dr. Johnson, is not that it took us so long to learn how to do this stuff; the wonder is that we can do it at all.

And in fact things began to pick up a little in these middle Middle Ages.[26] Al-Khwarizmi was followed by other mathematicians of note operating in Muslim lands, both eastern and western. Thabit ibn Qurra, of the generation after al-Khwarizmi and also based in Baghdad, did notable work in mathematical astronomy and the pure theory of numbers. A century and a half later, Mohammed al-Jayyani of Cordova in Muslim Spain wrote the first treatise on spherical trigonometry. None of them made significant progress in algebra, though. In particular, none attempted to replicate Diophantus's great leap into literal symbolism. All spelled out their problems in words, words, words.

I am going to give detailed coverage to only one other mathematician from medieval Islam, partly because he is worth covering, but also as a bridge to the Europe of the early Renaissance, where things *really* begin to pick up.

§3.7　Omar Khayyam is best known in the West as the author of the *Rubaiyat*, a collection of four-line poems offering a highly personal view of life—a sort of death-haunted hedonism with an alcoholic thread, somewhat prefiguring A. E. Housman. Edward Fitzgerald turned 75 of these into English quatrains, each rhymed *a-a-b-a*, in a translation published in 1859, and Fitzgerald's *The Rubaiyat of Omar Khayyam* was a great favorite all over the English-speaking world up to World War I. (An elaborate jeweled copy of the original went down with the *Titanic*.)

The *DSB* gives 1048–1131 as the most probable dates for Khayyam, and those are the dates I shall use *faute de mieux*. This puts Khayyam at least 250 years after al-Khwarizmi. It is worth bearing in mind these great gulfs of time when surveying intellectual activity in the Middle Ages.

The region in which Khayyam lived and worked was at the eastern end of the first great zone of Islamic conquest. That eastern-most region included Mesopotamia, the northern part of present-day Iran, and the southern part of Central Asia (present-day Turkmenistan, Uzbekistan, Tajikistan, and Afghanistan). In Khayyam's time this was a region of both ethnic and religious conflict. The principal ethnies involved were the Persians, the Arabs, and the Turks. The religious conflict was all within Islam: first between Sunnis and Shias, then between two factions of Shias, the main body and the split-off Ismailites.[27]

The Turks, originally nomads from farther Central Asia, had been hired as mercenaries by the declining Abbasids. Of course, the Turks soon realized what the true balance of power in the relationship was, and the later Abbasid caliphs, excepting a short-lived revival in the late 9th century, were puppets of their Turkish guards. Their only consolation was that the Turks had at least converted to orthodox (that is, Sunni) Islam. The farther eastern territories, beyond Mesopotamia, were anyway lost by the Abbasids to a Persian (and Shiite) revival in the 10th century. These short-lived Persian dynasties hired Turkish troops just as the Abbasids had. In due course a

Turkish general overthrew his Persian masters, establishing the Ghaznavid dynasty—the first Turkish empire. Wisely deferring to the now-subjugated Persians in matters of statecraft and high culture, the Ghaznavids ran a decent court, adorned with famous Persian scholars and poets. They also conducted several invasions of south Asia, carrying Islam to the region occupied by present-day Pakistan and India.

In 1037, a few years before Omar Khayyam was born, a man named Seljuk, one of the Ghaznavids' own Turkish mercenaries, rebelled and defeated the Ghaznavid armies. This new Turkish power expanded very fast. In 1055, when Omar was seven years old, Seljuk's grandson took Baghdad and gave himself the title of Sultan, meaning "ruler." The implication here was that the Caliph's power was now to be merely spiritual, like the Pope's.

The Seljuks ruled all the eastern territories of Islam through the remainder of the 11th century and much of the 12th. Their dominions extended all the way westward to the Holy Land and the borders of Egypt (at this time under Shiite rulers, of the Ismailite subpersuasion—the Seljuks were orthodox Sunnis). It was the Seljuk defeat of the Byzantine Emperor at the battle of Manzikert in 1071 that won them Anatolia, laying the first foundations of modern Turkey. It was the loss of Anatolia that caused the Byzantines to call on Western Europe for aid, thus precipitating the Crusades. And it was Seljuk Turks that the Crusaders faced on their trek across Anatolia to the Holy Land and at Antioch and Jerusalem.

§3.8   Omar Khayyam's life was therefore spent under the rule of the Seljuk Turks. His great patron was the third Seljuk sultan, Malik Shah, who ruled from 1073 to 1092 from his capital city of Esfahan in present-day Iran, 440 miles east of Baghdad.[28] Malik Shah is less famous than his vizier Nizam al-Mulk, one of the greatest names in the history of statecraft, a genius of diplomacy. Al-Mulk, like Khayyam, was a Persian. The two of them are sometimes spoken of together with Hasan Sabbah, founder of the Assassins sect,[29] as "the three Persians" of their time—that is, the three important men of the Persian ethny in the Seljuk empire.

Malik Shah's court seems to have been easygoing in matters of religion, as such things went in medieval Islam. This probably suited Khayyam very well. His poems show a skeptical and agnostic attitude to life, and his contemporaries often spoke of him as a freethinker. Anxious mainly to get on with his work—he was director of the great observatory as Esfahan—and studies, Khayyam did his best to stay out of trouble, writing orthodox religious tracts to order and probably performing the pilgrimage to Mecca that is every Muslim's duty. From the poems and such biographical facts as are available, Khayyam strikes the modern reader as rather *simpatico*.

His main interest for algebraists is a book he wrote in his 20s before going to Esfahan. The book's title is *Risala fi'l-barahin 'ala masa'il al-jabr wa'l-muqabala*, or "On the Demonstration of Problems in Completion and Reduction."

Like al-Khwarizmi and all the other medieval Muslim mathematicians, Khayyam ignores, or was altogether ignorant of, Diophantus's great breakthrough into literal symbolism. He spells out everything in words. Also, like the older Greeks, he has a strongly geometric approach, turning naturally to geometric methods for the solution of numerical problems.

Khayyam's main importance for the development of algebra is that he opened the first serious assault on the cubic equation. Lacking any proper symbolism and apparently unwilling to take negative numbers seriously, Khayyam was laboring under severe handicaps. The equation that we would write as $x^3 + ax = b$, for example, was expressed by Khayyam as: "A cube and a number of sides are equal to a number." Nonetheless he posed and solved several problems involving cubic equations, though his solutions were always geometrical.

This is not quite the first appearance of the cubic equation in history. Diophantus had tackled some, as we have seen. Even before that, Archimedes had bumped up against cubics when deliberating on such problems as the division of a sphere into two parts whose volumes have some given ratio.[30] (The connection with Archimedes' interest in floating bodies will occur to you if you think about this for a minute.) Khayyam seems to have been the first to recognize cubic equations as a distinct class of problems, though, and offered a classification of them into 14 types, of which he knew how to solve four by geometrical means.

As an example of the kind of problem Khayyam reduced to a cubic equation, consider the following:

> Draw a right-angled triangle. Construct the perpendicular from the right angle to the hypotenuse. If the length of this perpendicular plus the length of the triangle's shortest side equals the length of the hypotenuse, what can you say about the shape of the triangle?

The answer is that the ratio of the triangle's shortest side to the next shortest—a ratio that completely determines the triangle's shape—must satisfy the cubic equation

$$2x^3 - 2x^2 + 2x - 1 = 0$$

The only real-number solution of this equation is $0.647798871\ldots$, an irrational quantity very close to the rational number $\frac{103}{159}$. So a right-angled triangle with short sides 103 and 159 very nearly fills the bill, as the reader can easily verify.[31] Khayyam took an indirect approach, ending up with a slightly different cubic, which he solved numerically via the intersection of two classic geometric curves.

§3.9   To offer an extremely brief summary of events to date:

- The ancient Babylonians developed some techniques for solving a limited range of linear and quadratic equations with one unknown.

- The later Ancient Greeks tackled similar equations geometrically.
- Diophantus, in the 3rd century, broadened the scope of inquiry to many other kinds of equations, including equations
  of higher degree, equations with many variables, and systems
  of simultaneous equations. He also developed the first literal
  symbolism for algebraic problems.
- Medieval Islamic scholars gave us the word "algebra." They
  began to focus on equations as worthwhile objects of inquiry
  in themselves and classified linear, quadratic, and cubic equations according to how difficult it was to solve them with the
  techniques available.

In discussing Omar Khayyam, I mentioned the terrible battle of
Manzikert, the great retreat of Eastern Christendom that followed it,
and the strange, disorderly, and still controversial reaction to that retreat: the Crusades. By the time of these events—Khayyam's time—
Western European culture was beginning to struggle to its feet after
the Dark Ages. The lights came on earliest and brightest in Italy, and
it is there that we meet our next few algebraists.

Math Primer

CUBIC AND QUARTIC EQUATIONS

§CQ.1   THE FIRST GREAT ADVANCE in algebra after the Middle Ages was the general solution of the cubic equation, immediately followed by a solution for the quartic. I tell the full story in the next chapter. There, however, I shall be looking at the topic through 16th-century eyes. Here I just want to give a brief modern account of the underlying algebra to clarify the issues and difficulties.

It is important to understand what is being sought here. It is not difficult to find approximate *numerical* solutions to cubic equations, or equations of any degree, to any desired accuracy. Sometimes you can just guess a correct solution. Drawing a graph will often get you very close. More sophisticated arithmetic and geometric methods were known to the Greeks, Arabs, and Chinese. Medieval European mathematicians were familiar with these methods too and could generally figure out a numerical value for the real solution of a cubic equation to good accuracy.

What they did not have was a properly *algebraic* solution—a universal formula for the solutions of a cubic equation, like the one for the quadratic equation in Endnote 14. A formula of that general kind was what early-modern mathematicians sought. Only when it was found could the cubic equation be considered solved.

§CQ.2   Here is a general cubic equation[32] in one unknown: $x^3 + Px^2 + Qx + R = 0$. The first thing we can do is drop the term in x^2, just by noticing the following simple algebraic fact:

$$x^3 + Px^2 + Qx + R = \left(x + \frac{P}{3}\right)^3 + \left(Q - \frac{P^2}{3}\right)\left(x + \frac{P}{3}\right) + \left(R - \frac{QP}{3} + \frac{2P^3}{27}\right)$$

To put it another way,

$$x^3 + Px^2 + Qx + R = X^3 + \left(Q - \frac{P^2}{3}\right)X + \left(R - \frac{QP}{3} + \frac{2P^3}{27}\right),$$

where $X = x + P/3$. So by dint of a simple substitution, any cubic equation can be converted to one with no x^2 term, and if we can solve the simpler equation by finding X, we can then easily get x by simply subtracting $P/3$. This kind of cubic, one with no x^2 term, is technically known by the rather charming name *depressed cubic*.[33]

The long and short of it is that we need only bother with depressed cubic equations, ones that look like this:

$$x^3 + px + q = 0$$

§CQ.3   So far, so good, but what is the general solution of that depressed cubic? Cast your mind back to the general quadratic equation

$$x^2 + px + q = 0$$

with its two solutions

$$x = \frac{-p + \sqrt{p^2 - 4q}}{2} \quad \text{and} \quad x = \frac{-p - \sqrt{p^2 - 4q}}{2}$$

Since negative numbers do not have real-number square roots, if $p^2 - 4q$ is negative, the solutions are complex numbers. If $p^2 - 4q$ is precisely zero, then the two solutions are the same; so from a numerical point of view there is only one solution. And, of course, if $p^2 - 4q$ is positive, there are two real-number solutions.

All of this can easily be illustrated by plotting graphs of quadratic polynomials. If we plot the graph $y = x^2 + px + q$, the solutions of the quadratic equation are where $y = 0$, that is, where the curve cuts the horizontal axis. The three cases I covered above—no real solutions, one real solution, two real solutions—are shown in Figures CQ-1, CQ-2, and CQ-3.

To see what we might expect from the cubic, we could *start* from the graphical picture. There are three basic situations, illustrated in Figures CQ-4, CQ-5, and CQ-6. Notice that in all three cubics the curve starts in the far southwest and ends up in the far northeast. This is because when x is very big (positive or negative), the x^3 term in $x^3 + px + q$ "swamps" the other terms. For "big enough" values of x, $x^3 + px + q$ "looks like" x^3. (The precise size involved in "big enough" depends on how big p and q are.) Now, the cube of a negative num-

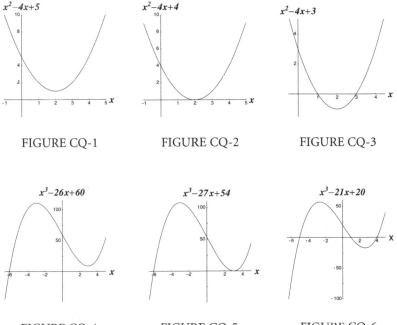

FIGURE CQ-1	FIGURE CQ-2	FIGURE CQ-3

FIGURE CQ-4	FIGURE CQ-5	FIGURE CQ-6

ber, by the rule of signs, is negative, and the cube of a positive number is positive. So the graph of a cubic polynomial will always go southwest to northeast. It follows, since the curve must cross the horizontal axis *somewhere*, that there must always be *at least one* real solution to $x^3 + px + q = 0$.

As the graphs show, in fact, there can be one, two, or three solutions of a cubic equation, but there cannot be none, and there cannot be more than three. On the basis of our experience with the quadratic equation, you might suspect that, in most of the cases where there is only one real solution, there are two complex solutions hidden away out of sight. That is correct.

§CQ.4   Here, in fact, are the general three solutions of the cubic equation $x^3 + px + q = 0$:

$$x = \sqrt[3]{\frac{-q+\sqrt{q^2+\left(4p^3/27\right)}}{2}} + \sqrt[3]{\frac{-q-\sqrt{q^2+\left(4p^3/27\right)}}{2}}$$

$$x = \left(\frac{-1+i\sqrt{3}}{2}\right)\sqrt[3]{\frac{-q+\sqrt{q^2+\left(4p^3/27\right)}}{2}} + \left(\frac{-1-i\sqrt{3}}{2}\right)\sqrt[3]{\frac{-q-\sqrt{q^2+\left(4p^3/27\right)}}{2}}$$

$$x = \left(\frac{-1-i\sqrt{3}}{2}\right)\sqrt[3]{\frac{-q+\sqrt{q^2+\left(4p^3/27\right)}}{2}} + \left(\frac{-1+i\sqrt{3}}{2}\right)\sqrt[3]{\frac{-q-\sqrt{q^2+\left(4p^3/27\right)}}{2}}$$

That needs a little explanation. In fact, it needs a *lot* of explanation.

First note that the two things under the cube root on each line differ by only a sign, plus or minus. If you look more closely, in fact, you will see that they are the two solutions of the quadratic equation

$$t^2 + qt - \frac{p^3}{27} = 0$$

(I switched from x to t as my "dummy variable" to avoid confusion with the main topic.) Let's call these two numbers α and β for convenience.

Next, what are those complex numbers in the second and third lines? Well, they are cube roots of 1. In the world of real numbers, of course, the number 1 has only one cube root, namely itself, since $1\times1\times1 = 1$. In the world of complex numbers, however, it has two more, and there they are. If you glance forward to §RU.1 in my primer on the roots of unity, between Chapters 6 and 7, you will see that the first of these complex numbers, as they appear above, is commonly denoted by ω, the second by ω^2, and that each is the square of the other.

Now I can write down the solutions of the general cubic equation very succinctly. They are

$$x = \sqrt[3]{\alpha} + \sqrt[3]{\beta},$$

$$x = \omega\sqrt[3]{\alpha} + \omega^2\sqrt[3]{\beta},$$

$$x = \omega^2\sqrt[3]{\alpha} + \omega\sqrt[3]{\beta},$$

where ω and ω^2 are the complex cube roots of 1, and α and β are the solutions of the quadratic equation $t^2 + qt - (p^3/27) = 0$.

§CQ.5   Just looking at the three solutions like that, it seems that the first is a real number and that the second and third are complex numbers. How can that be, since we know that a cubic equation can have three real solutions (Figure CQ-6)?

The snag is that both α and β may be complex numbers themselves. Both of them contain the square root of $q^2 + (4p^3/27)$, and this might be negative. If it is, then α and β are both complex; if it is not, α and β are both real. The case where they are both complex is the so-called *irreducible case*.[34] Then you are stuck with finding the cube root of a complex number—no easy matter. This makes the general solution of the cubic, while intellectually very satisfying, not very practical.

The actual status of the three solutions, in relation to the sign of $q^2 + (4p^3/27)$, which I am going to refer to (rather loosely, I am afraid) as the *discriminant* of the equation, is summarized in the following table. The bottom row represents the irreducible case.

	1st solution	*2nd solution*	*3rd solution*
Discriminant positive	Real	Complex	Complex
Discriminant zero	Real	Real and equal	
Discriminant negative	Real	Real	Real

If the discriminant is zero, then α and β are equal; so the three solutions boil down to $2\sqrt[3]{\alpha}$, $(\omega + \omega^2)\sqrt[3]{\alpha}$, $(\omega^2 + \omega)\sqrt[3]{\alpha}$. The second two are of course equal, and $(\omega + \omega^2)$, as you can very easily verify, is -1.

§CQ.6   I gave the solution of the general cubic without a proof. The proof, with modern symbolism, is not difficult. To solve $x^3 + px + q = 0$, first express x as the sum of two numbers: $x = u + v$. This can, of course, be done in an infinity of ways. The cubic equation now looks like this: $(u + v)^3 + p(u + v) + q = 0$. This can be rearranged to the following:

$$(u^3 + v^3) + 3uv(u + v) = -q - p(u + v)$$

So if, from the infinity of possibilities for u and v, I pick the particular values such that

$$u^3 + v^3 = -q \text{ and } 3uv = -p,$$

then I shall have a solution. Extracting v in terms of u and p from the second of those equations, and feeding it back into the first, I get this:

$$u^6 + qu^3 - \frac{p^3}{27} = 0$$

That is just a quadratic equation in the unknown u^3. Since I could just as easily have extracted u instead of v, getting the identical equation with v as variable, u^3 and v^3 are the solutions of this quadratic equation. Then u and v are the cube roots of those solutions, or the cube roots multiplied by a cube root of 1. So possible solutions are $u + v$, $\omega u + \omega^2 v$, and $\omega^2 u + \omega v$. (Note that combinations like $u + \omega v$ are *not* possible because of the condition $3uv = -p$, which means that multiplying together the numbers on each side of the plus sign must yield a real number. ωu times $\omega^2 v$ is a real number when uv is, because $\omega^3 = 1$)

§CQ.7   The solution of the general quartic equation

$$ax^4 + bx^3 + cx^2 + fx + g = 0$$

is something of an anticlimax after all that. As before, we usually simplify, or "depress," the equation to this form:

$$x^4 + px^2 + qx + r = 0$$

By some straightforward substitutions, this can be rewritten as a difference of two squared expressions. In solving those, you end up with a cubic equation: The quartic reduces to a cubic in much the same way the cubic reduces to a quadratic.

The four solutions are expressions in square and cube roots involving p, q, and r. To write all four out in full would take more space than I can justify, but here is one of them:

$$x = \frac{1}{2}\sqrt{-\frac{2p}{3}+\frac{\sqrt[3]{2}\,t}{3\sqrt[3]{u}}+\frac{\sqrt[3]{u}}{3\sqrt[3]{2}}}$$

$$-\frac{1}{2}\sqrt{-\frac{4p}{3}-\frac{\sqrt[3]{2}\,t}{3\sqrt[3]{u}}-\frac{\sqrt[3]{u}}{3\sqrt[3]{2}}-\frac{2q}{\sqrt{-\frac{2p}{3}+\frac{\sqrt[3]{2}\,t}{3\sqrt[3]{u}}+\frac{\sqrt[3]{u}}{3\sqrt[3]{2}}}}},$$

where $t = p^2 + 12r$ and

$$u = 2p^3 + 27q^2 - 72pr + \sqrt{\left(2p^3 + 27q^2 - 72pr\right)^2 - 4t^3}$$

This is fearsome looking, but it's just arithmetic, with square and cube roots and combinations of p, q, and r. Given the step up in complexity between the general solution of the quadratic equation and the general solution of the cubic, this is another step up of similar magnitude.

You might suppose that a general solution for the *quintic* equation—that is, an equation of the fifth degree—would be more complicated yet, perhaps filling a page or more, but that it would consist of nothing but square roots, cube roots, and probably fifth roots, though perhaps nested inside each other in various ways. Given our experience with the quadratic, cubic, and quartic equations, this is an entirely reasonable supposition. Alas, it is wrong.

Chapter 4

COMMERCE AND COMPETITION

§4.1 A PLEASING LITTLE CURIOSITY in historical writing is *The Thirteenth, Greatest of Centuries* by James J. Walsh, published in 1907. Much of the book consists of Roman Catholic apologetics (Walsh was a professor at Fordham University, a Jesuit foundation in New York City), but the author makes a good case that the 13th century has, at the very least, been much underestimated as one of progress, cultural achievement, and the recovery of classical learning. The great gothic cathedrals; the early universities; Cimabue and Giotto; St. Francis and Aquinas; Dante (just) and the *Romance of the Rose*; Louis IX, Edward I, and Frederick II; Magna Carta and the Guilds; Marco Polo and Friar Odoric (who seems to have reached Lhasa). . . . There was a great deal going on in the 13th century. It was in the early decades of that century that Leonardo of Pisa, better known as Fibonacci, flourished.

Fibonacci's is one of the mathematical names best known to nonmathematicians, because of the much-publicized Fibonacci sequence:

$$1, 1, 2, 3, 5, 8, 13, 21, 34, 55, 89, 144, 233, 377, \ldots$$

Each term in this sequence is the sum of the two to its left: $89 = 34 + 55$. This sequence—it is number A000045 in the addictive *Online Encyclopedia of Integer Sequences*—has so many mathematical and scientific connotations that there is a journal devoted to it: the *Fibonacci Quarterly*. The August 2005 issue includes articles with titles such as: "*p*-Adic Interpolation of the Fibonacci Sequence via Hypergeometric Functions."

It is in fact quite easy, though a bit surprising to the nonmathematician, to show[35] that the *n*th term of the sequence is precisely

$$\frac{1}{\sqrt{5}}\left[\left(\frac{1+\sqrt{5}}{2}\right)^n - \left(\frac{1-\sqrt{5}}{2}\right)^n\right]$$

If *n* is equal to 4, for example, this works out, using the binomial theorem,[36] to

$$\frac{1}{\sqrt{5}}\left[\left(\frac{1+4\times\sqrt{5}+6\times 5+4\times 5\sqrt{5}+25}{16}\right) - \left(\frac{1-4\times\sqrt{5}+6\times 5-4\times 5\sqrt{5}+25}{16}\right)\right]$$

which can easily be seen to be equal to 3.

The Fibonacci sequence first appeared in a book, *Liber abbaci,* written by Leonardo of Pisa.[37] The context is a number problem about rabbits.

> How many pairs of rabbits will be produced in a year, beginning with a single pair, if in every month each pair bears a new pair which becomes productive from the second month on, and no deaths occur?

The only way I have ever been able to think this problem through is to label the months *A, B, C, D,* etc. The rabbit pair we start with is the *A* pair. In the second month we still have the *A* pair but also the *AB* pair they have begotten, for a total two pairs. In the third month the *A* pair begets another pair, the *AC* pair. The *AB* pair is present but not yet begetting. Total pairs: three. In the fourth month, month *D,* the *A* pair is still with us and has begotten another pair, the *AD* pair.

The *AB* pair is also with us and has now begotten a pair of its own, the *ABD* pair. The *AC* pair is present but not yet begetting. *A*, *AD*, *AB*, *ABD*, *AC*—a total of five pairs . . . and so on.

§4.2 In the preface to *Liber abbaci*, the author gives us some details of his life up to that point. The modern style of given names plus surname had not yet "settled down" in Western Europe, so the author is formally known only as Leonardo of Pisa, sometimes Italianized to Leonardo Pisano. He belonged to a clan that did have a name for itself, the Bonacci clan, and so he got tagged, or tagged himself, with *fi'Bonacci*, "son of the Bonaccis," and it stuck.

Leonardo was born around 1170 in Pisa, hometown of the Bonaccis.[38] Pisa, though surrounded by the territories of the German—that is, the Holy Roman—empire, was an independent republic at this point. Leonardo's father was an official of that republic, and around 1192 he was appointed to represent the merchants in the Pisan trading colony of Bugia,[39] on the North African coast. Soon afterward he sent for Leonardo to join him. The idea was that the young man would train to be a merchant.

Bugia was planted on Islamic territory, this stretch of North Africa—as well as the southern third of Spain—being at the time under the rule of a Shi'ite dynasty, the Muwahids (also written "Almohads"), who ruled from Marrakech in the far west. The young Leonardo was thus exposed to all the learning that might be available in a large Muslim city, presumably including the works of medieval Islamic mathematicians such as al-Khwarizmi and Omar Khayyam. His father soon sent him off on business trips all over the Mediterranean—to Egypt and Syria, Sicily (a Norman kingdom until 1194, when the Hohenstaufen dynasty of Germany inherited it), France, and Byzantium.

There must have been many other young men from the trading cities of Italy engaged in similar travels. Leonardo, however, was a born mathematician, and in his travels he skimmed off the best that was known at that time from the Byzantine Greeks and the Muslims—and, via the Muslims, from Persia, India, and China. When he returned to settle permanently in Pisa around the year 1200, he probably had a wider knowledge of arithmetic and algebra, as those disciplines existed in his time, than anyone in Western Europe—perhaps anyone in the world.

§4.3   *Liber abbaci* was, by the standards of its time, wonderfully innovative and very influential. For 300 years it was the best math textbook available that had been written since the end of the ancient world. It is often credited with having introduced "Arabic" (that is, Indian) numerals, including zero, to the West. The book begins, in fact, with this:

> These are the nine figures of the Indians: 9 8 7 6 5 4 3 2 1. With these
> nine figures, and with the sign 0 which in Arabic is called *zephirum*,
> any number can be written, as will be demonstrated.

The first 7 of the book's 15 chapters are a primer in computation using these "new" numerals, with many worked examples. The remainder of the book is a collection of problems in arithmetic, algebra, and geometry, some of a type to interest merchants and artisans, some lighthearted recreational puzzles like the rabbit problem described earlier, which appears in Chapter 12 of the book.

While fascinating arithmetically and historically, *Liber abbaci* is not the most *algebraically* interesting of Fibonacci's works. Two later books, written probably in 1225, show his algebraic skills to best effect. I shall pick just the first one because it leads to the main topic of the rest of this chapter: the cubic equation.

§4.4   In or about 1225 the German emperor held court in Pisa. This emperor was Frederick II, one of the most fascinating characters of his age, sometimes called the first modern man to sit on a European throne. There is a good brief character sketch of him in Volume III of Steven Runciman's *History of the Crusades*. Frederick enjoyed the distinction (one cannot help thinking that he *did* enjoy it) of having been excommunicated twice by the Pope himself, as part of the power game between pope and emperor that raged during this period. He has remained unpopular with Roman Catholics ever since. The aforementioned Professor Walsh, in his paean to the 13th century, mentions Frederick, one of the most intelligent and cultivated rulers of that century, just once in 429 pages.

By this date, Fibonacci's mathematical talents were well known in Pisa, and he was also on friendly terms with some of the scholars at Frederick's court. So when Frederick came to Pisa, Leonardo was granted an audience.

Frederick had in his court one Johannes of Palermo (a.k.a. Giovanni da Palermo, John of Palermo), a person about whom I have been able to discover very little. One source says he was a Marrano, that is to say a Spanish Jew who had converted to Christianity. At any rate, Johannes seems to have been knowledgeable in math, and Frederick asked him to set some problems for Fibonacci, to test the man's abilities.

One of the problems was to solve the cubic equation $x^3 + 2x^2 + 10x = 20$. Whether Fibonacci was able to solve this problem on the spot, I do not know. At any rate, he wrote it up in one of those books he issued in 1225, a book with a lengthy Latin title generally shortened to *Flos*.[40] The actual real solution of this equation is 1.3688081078213726. . . . Fibonacci gave a result very close to this— wrong only in the 11th decimal place.

We don't know how Fibonacci got his result. He doesn't tell us. Probably he used a geometric method, like the intersecting curves Omar Khayyam employed for the same purpose. What is notable about his treatment of this cubic is in fact not his solution of it, but his analysis. He first shows, by meticulous reasoning, that the solution cannot be a whole number. Then he shows that it cannot be a rational number either. Then he shows that it cannot be a square root, or any combination of rational numbers and square roots. This analysis of a cubic is a *tour de force* of medieval algebra. In its concentration on the *nature* of the solution rather than its actual value, it might be said to have anticipated the great revolution in thinking about the solutions of equations that took place 600 years later, which I shall describe in due course.

§4.5   Fibonacci did not express his solution of that cubic in our familiar decimal form, but rather in sexagesimal, like an ancient Babylonian, as $1° 22' 7'' 42''' 33^{IV} 4^{V} 40^{VI}$. This means

$$1 + (22 + (7 + (42 + (33 + (4 + 40 \div 60) \div 60) \div 60) \div 60) \div 60) \div 60,$$

which works out to 1.3688081078532235. . . . This is, as I said earlier, correct to 10 decimal places. The great obstacle to the development of algebra in late medieval times was in fact the absence of good ways to write down numbers, unknown quantities, and arithmetic operations. Fibonacci's popularization of Hindu-Arabic numerals in Europe was a great advance, but until true decimal positional notation was applied to the fractional part of a number, too, this first "digital revolution" was not complete.

In the matter of expressing the unknown quantity and its powers, the situation was even worse. Outside purely geometrical demonstrations, the Muslim algebraists had, as I have said, done everything with words, commonly using the Arabic words *shai* ("thing") or *jizr* ("root") to stand for the unknown, with *mal* ("wealth" or "property") for the square of the unknown, *kab* ("cube") for its cube, and combined forms for higher powers: *mal-mal-shai* for the fifth power, and so on. Knowledge of Diophantus's much snappier notation was apparently preserved in the Greek libraries at Constantinople,[41] but mathematicians in the Muslim and Western-Christian worlds seem not to have been aware of it or did not feel the need for it.

Early Italian algebraists such as Fibonacci followed the Muslims, translating their words into Latin or Italian: *radix* ("root"), *res, causa,* or *cosa* ("thing"), *census* ("property"), *cubus* ("cube"). By the later 14th century these latter three were being abbreviated to *co, ce,* and *cu,* a development that was systematized in a book published in 1494 by an Italian named Luca Pacioli and commonly called Pacioli's *Summa*.[42] Pacioli's notation was wider in scope than Diophantus's ς, Δ^Y, and K^Y, but less imaginative. Though there is little original work in the *Summa*, it proved very handy for commercial arithmeticians and enduringly popular. Pacioli is considered the father of double-entry bookkeeping.[43]

§4.6   I skipped rather blithely across 269 years there without saying anything about what happened in the interim. This was in part authorial license: I want to get to the solution of the general cubic, and to Cardano, the first real personality in my book. It was also, though, because nothing of much note *did* happen between Fibonacci and Pacioli.

There were certainly algebraists at work in the 13th, 14th, and early 15th centuries. More technical histories of algebra list some of their contributions. Van der Waerden, for example, gives nearly six pages to Maestro Dardi of Pisa, who tackled quadratic, cubic, and quartic equations in the middle of the 14th century, classified them into 198 types, and used ingenious methods to solve particular types.

While noteworthy to the specialist, these secondary figures added little to what was understood. It was only with the spread of printed books during the second half of the 15th century that the development of algebra really picked up speed.

By no means was all of the action in Italy. The Frenchman Nicolas Chuquet produced a manuscript (it was not actually printed until 1880) titled *Triparty en la science des nombres* in 1484, introducing the use of superscripts for powers of the unknown (though not quite in our style: he wrote 12^3 for $12x^3$) and treating negative numbers as entities in themselves. The German Johannes Widman gave the first lecture on algebra in Germany (Leipzig, 1486) and was the first to use the modern plus and minus signs in a printed book, published in 1489.[44]

Except for Chuquet's superscripts, which were little noticed, all of this work still clung to the late medieval style of notation for the unknown and its powers, the unknown itself being *chose* in French or *coss* in German.[45] Nor were there any very significant discoveries until the solution of the general cubic equation around 1540. It is to that fascinating story that I now turn.

§4.7   At the center of the story is Girolamo Cardano, who was born at Pavia in 1501, died in Rome in 1576, but was raised and spent most of his life in or near Milan, which he considered his hometown.

Cardano is a large and fascinating personality, "a piece of work," we might say nowadays. Several biographies of him have been written, the first by himself: *De Propria Vita*, which he produced near the end of his life. This autobiography contains a list, covering several pages, of his other books. He counts 131 printed works, 111 unprinted books in manuscript form, and 170 manuscripts he claims to have destroyed as unsatisfactory.

Many of these books were Europe-wide best sellers. We know, for example, that *Consolation*, his book of advice to the sorrowing, first translated into English in 1573, was read by William Shakespeare. The sentiments in Hamlet's famous "To be, or not to be" soliloquy closely resemble some remarks about sleep in *Consolation*, and this may be the book that Hamlet is traditionally carrying when he comes on stage to deliver that soliloquy.

Cardano's first and main interest, and the source of his livelihood, was medicine. His first published book was also about medicine, offering some commonsense remedies and mocking some of the stranger, positively harmful, medical practices of the time.

(Cardano claimed that he wrote the book in two weeks.) By the time he reached 50, Cardano was the second most famous physician in Europe, after Andreas Vesalius. The high society of the time, both lay and clerical, clamored for his services. He seems to have been averse to travel, however, only once venturing far afield—to Scotland in 1552, to cure the asthma of John Hamilton, the last Roman Catholic archbishop of that country. Cardano's fee was 2,000 gold crowns. The cure seems to have been completely successful: Hamilton lived until 1571, when he was hanged, in full pontificals, on the public gibbet at Stirling for complicity in the murder of Lord Darnley, husband of Mary Queen of Scots.

Before the advent of copyright laws, writing books, even best sellers, was not a path to wealth, except indirectly, by way of self-advertisement. Cardano's main secondary sources of income were from gambling and the casting of horoscopes. Passing through London on his way home from Scotland, Cardano cast the horoscope of the boy king Edward VI (Henry VIII's son), predicting a long life despite illnesses the king would suffer at ages 23, 34, and 55. Unfortunately Edward died less than one year later at age 16. Other forms of divination also got Cardano's attention. He even claimed to have invented one: "metoposcopy," the reading of character and fate from facial irregularities. A sample from Cardano's book on this subject: "A woman with a wart upon her left cheek, a little to the left of the dimple, will eventually be poisoned by her husband."

Cardano's attraction to gambling probably rose to the level of an addiction and might have ruined him but for the fact that he was a keenly analytical chess player—chess in those days being commonly played for money—and possessed a superior understanding of mathematical probability. He wrote a book about gambling, *Liber de ludo aleae* ("A Book About Games of Chance"), containing some careful mathematical analyses of dice and card games.[46]

In the true Renaissance spirit, Cardano excelled in practical sciences as well as theoretical ones. His books are rich in pictures of devices, mechanisms, instruments, and methods for raising sunken ships or measuring distance. When Holy Roman Emperor Charles V came to Milan in 1548, Cardano had a place of honor in his procession, having designed a suspension device for the emperor's carriage. (Charles suffered badly from gout and did not enjoy traveling—an unfortunate thing in a man whose European dominions stretched from the Atlantic to the Baltic.[47]) The universal joint used in automobiles today is still named after Cardano in French (*le cardan*) and German (*das Kardangelenk*).

The lowest point of Cardano's long life was the execution of his son Giambatista, whom he adored and in whom he had invested great hopes. The boy fell in love with a worthless woman and married her. After she had borne three children and taunted Giambatista that none of them were his, he poisoned her with arsenic. (Being poisoned by one's husband seems to have been an occupational hazard for 16th-century Italian wives.) Quickly arrested, Giambatista was tortured and mutilated before execution. He was not quite 26 years old. This dreadful event haunted the remaining 16 years of Cardano's life. Then, near the very end of that life, Cardano himself was imprisoned for heresy by the authorities of the Counter-Reformation. We don't know the charges against him. In his autobiography he does not tell us; presumably he was sworn to silence. After a few months in jail he was released to house arrest, but he was no longer permitted to lecture publicly or to have books published.

For all his adventures and misfortunes, Cardano died peacefully in his bed on September 20, 1576, nearly 75 years old. This was precisely the date he had predicted when casting his own horoscope some years earlier. There were those who said he poisoned himself, or starved to death, just to make the date come out right. It would not have been out of character.

§4.8   Cardano's prominence in the history of algebra rests on his book *Artis magnae sive de regulis algebraicis liber unus*—"Of the Great Art, or the First Book on the Rules of Algebra." This work contains the general solution of the cubic and quartic equations and also the first serious appearance of complex numbers in mathematical literature. *Ars magna*, as the book is always called, was first printed in Nuremberg in 1545.

Luca Pacioli, in the *Summa*, had listed two types of cubics as having no possible solution:

$$(1) \quad n = ax + bx^3$$
$$(2) \quad n = ax^2 + bx^3$$

These were known as, respectively, "the cosa and the cube equal to a number" and "the censi and the cube equal to a number." A third type, not listed by Pacioli as impossible (I don't know why) was "the cosa and a number equal to a cube":

$$(3) \quad ax + n = bx^3$$

This looks to be the same as a type 1 to us, but that's because we take negative numbers in our stride. In Cardano's time, negative numbers were only just beginning to be acknowledged as having independent existence.

At some point in the early 16th century, a person named Scipione del Ferro found the general solution to the type-1 cubic. Del Ferro was professor of mathematics at the University of Bologna; his dates are ca. 1456–1526. We don't know exactly when he got his solution or whether he also solved type 2. He never published his solution.

Before del Ferro died, he imparted the secret of his solution for "the cosa and the cube" to one of his students, a Venetian named Antonio Maria Fiore. This poor fellow has gone down in all the history books as a mediocre mathematician. I don't doubt the judgment of the historians, but it seems a great misfortune for Fiore to have gotten mixed up—as a catalyst, so to speak—in such a great and algebraically critical affair, so that his mathematical mediocrity echoes down the ages like this. At any rate, having gotten the secret of the cosa and the cube, he decided to make some money out of it. This wasn't hard to do in the buzzing intellectual vitality of northern Italy at the time. Patronage was hard to come by, university positions were not well paid, and there was no system of tenure. For a scholar to make any kind of living, he needed to publicize himself, for example, by engaging in public contests with other scholars. If some large cash prize was at stake in the contest, so much better the publicity.

One mathematician who had made a name for himself in this kind of contest was Nicolo Tartaglia, a teacher in Venice. Tartaglia came from Brescia, 100 miles west of Venice. When he was 13, a French army sacked Brescia and put the townsfolk to the sword. Nicolo survived but suffered a grievous saber wound on his jaw, which left him with a speech impediment: Tartaglia means "stutterer"—this was still the age when last names were being formed out of locatives, patronymics, and nicknames. Tartaglia was a mathematician of some scope, author of a book on the mathematics of artillery, and the first person to translate Euclid's *Elements* into Italian.

In 1530, Tartaglia had exchanged some remarks about cubic equations with another native of Brescia, a person named Zuanne de Tonini da Coi, who taught mathematics in that town. In the course of those exchanges, Tartaglia claimed to have found a general rule for the solution of type-2 cubics, though he confessed he could not solve type 1.

Somehow Fiore, the mathematical mediocrity, heard of these exchanges and of Tartaglia's claim. Either believing Tartaglia to be bluffing or confident that he was the only person who knew how to solve type-1 cubics (the secret he had gotten from del Ferro), Fiore challenged Tartaglia to a contest. Each was to present the other with 30 problems. Each was to deliver the 30 solutions to the other's problems to a notary on February 22, 1535. The loser was to stand the winner 30 banquets.

Having no great regard for Fiore's mathematical talents, Tartaglia at first did not bother to prepare for the contest. However, someone passed on the rumor that Fiore, though no great mathematician himself, had learned the secret of solving "the cosa and the cube" from a master mathematician, since deceased. Now worried, Tartaglia bent his talents to finding a general solution of type-1 cubics. In the small hours of the morning of Saturday, February 13, he cracked it. As he had suspected, all of Fiore's problems were type-1 cubics, the solution of which was Fiore's sole claim to mathematical ability.

Tartaglia's questions seem (we only have the first four) to have been a mix of types 2 and 3. It is plain that at this point Tartaglia had mastered all the cubics, of any type, having just one real solution—all the ones, that is, with a positive discriminant. Cubic equations with a negative discriminant (and therefore having three real solutions) can only be solved by manipulating complex numbers, which had not yet been discovered.

At any rate, Tartaglia was able to solve all of Fiore's problems, while Fiore could solve none of his. Tartaglia took the honor but waived the stake. Comments Cardano's biographer: "The prospect of thirty banquets face to face with a sad loser may have been rather uninspiring to him."[48]

§4.9 Cardano heard of Tartaglia's triumph from da Coi, that same native of Brescia with whom Tartaglia had exchanged remarks about cubic equations in 1530. Da Coi had moved to Milan after his exchanges with Tartaglia. Teachers of mathematics were not in very plentiful supply in northern Italy, and Cardano engaged da Coi to teach one of his classes. It seems to have been from da Coi that Cardano got a full account of the Fiore–Tartaglia duel and about the Tartaglia–da Coi exchanges of five years earlier. At this time Cardano was writing a book whose title he envisioned as *The Practice of Arithmetic, Geometry, and Algebra*. Probably he thought that Tartaglia's solution of the cubic, if he could get it, would go very nicely into the book. He accordingly embarked on a campaign to tease the secret out of Tartaglia.

The exchanges that followed make fascinating reading.[49] Cardano plays Tartaglia like a master angler reeling in a fish, alternating from haughty deprecation to sweet seduction, in a correspondence that lasted through January, February, and March of 1539. The choicest bait on Cardano's hook was the prospect of his introducing Tartaglia to Alfonso d'Avalos, one of the most powerful men in Italy, governor (that is, under Emperor Charles V) of all Lombardy, and commander of the imperial army stationed near Milan. Tartaglia's book on artillery had come out not long before, and Cardano claimed to have bought two copies, one for himself and one for his friend the governor. His Excellency (promised Cardano, with what truth we do not know) was anxious to meet the author.

Tartaglia hurried to Milan and stayed for several days at Cardano's house. To switch metaphors, the fly had made straight for the spider's web. The governor was unfortunately out of town, but Cardano treated his guest with royal hospitality, and Tartaglia finally yielded the secret of the cosa and the cube on March 25. He insisted, however, that Cardano swear a solemn oath never to reveal it. Cardano duly swore, and Tartaglia wrote down his solution to the cubic as a poem of 25 lines. The poem begins:

Quando che'l cubo con le cose appresso
Se agguaglia a qualche numero discreto . . .

(When the cube and the cosa together
Are equal to some whole number . . .)

Tartaglia, by his own account, suffered from (to switch metaphors yet again) post-seduction remorse as soon as he had left Cardano's house. He went home to Venice and brooded. Cardano wrote to ask for clarification of some points in the poem, but Tartaglia's response was brusque. He was mollified somewhat when Cardano's arithmetic book came out in May; his solution of the cubic did not appear in it. That summer, however, he heard that Cardano had started work on another book, to deal specifically with algebra. Some further exchanges followed—angry and suspicious on Tartaglia's part, soothing on Cardano's—into 1540.

Ars magna was published in 1545. The five years between that last exchange of letters early in 1540 and the publication of *Ars magna* were critical in the history of algebra. Cardano, having gotten the secret of the cosa and the cube, proceeded to a general solution of the cubic equation.

From studying the irreducible case, he came to realize that there must *always* be three solutions. To deal with this, of course, he had to come to terms with complex numbers. He did so hesitantly and incompletely, with many doubts, which should not surprise us. Even negative numbers were still thought of as slightly mysterious. Imaginary and complex numbers must have seemed positively occult. (They still do to many people.)

Here is Cardano in Chapter 37 of *Ars magna*, struggling with the following problem, which is quadratic, not cubic: Divide 10 into two parts whose product is 40.

> Putting aside the mental tortures involved, multiply $5+\sqrt{-15}$ by $5-\sqrt{-15}$, making $25-(-15)$, which [latter] is $+15$. Hence this product is 40. . . . This is truly sophisticated. . . .

Indeed it was. Cardano must have labored long and hard to make such a breakthrough. His ideas went off in other directions, too. He found some numerical methods for getting approximate solutions and formed ideas about the patterns of relationship between solutions and coefficients, thereby glimpsing territory that mathematicians did not begin to explore until 150 years later.

Cardano had help in his labors. Back in 1536, he had taken on a 14-year-old lad named Lodovico Ferrari as a servant. He found the boy unusually intelligent, already able to read and write, so he promoted him to the position of personal secretary. Ferrari learned math by proofing the manuscript for the 1540 arithmetic book. We can assume that when Cardano was wrestling with cubic equations, he shared his explorations with his young secretary.

One reason we can assume this is that in 1540, Ferrari worked out the solution to the general quartic equation. As I mentioned in my primer, this involves solving a cubic; so Ferrari could not publish his result without publishing the solution to the cubic, which he had learned from Cardano and which Cardano had sworn to Tartaglia he would not reveal.

Meanwhile, in the years since Scipione del Ferro's death in 1526 and the Fiore–Tartaglia duel in 1535, rumors had been going round that Fiore had gotten the solution to the cube and the cosa from the late del Ferro. Spotting a possible escape from their joint moral dilemma, in 1543, Cardano and his secretary Ferrari journeyed to Bologna to talk to del Ferro's successor at the university, who was also his son-in-law and custodian of his papers. After examining those papers, Cardano and Ferrari knew that Tartaglia had not been the first to solve the cube and the cosa. Thus supplied with a moral loophole, Cardano went ahead and included the full solutions to the cubic and quartic in *Ars magna*. He credited del Ferro as the one who first found a solution to the cube and the cosa and Tartaglia with having rediscovered it.

Tartaglia, who had spent the five years working quietly on his translations of Euclid and Archimedes, was of course furious. Three years of vituperative feuding followed, though Cardano kept out of it, leaving Ferrari to fight his corner. It all ended with another scholarly challenge-contest between Tartaglia and Ferrari in Milan, on August 10, 1548. We have only a brief and suspect account of the proceedings from Tartaglia's pen. It seems clear that he got the worst of the contest.

Tartaglia died in 1557, still angry and bitter. He never did publish the solution to the cubic himself, and no unpublished version was found among his papers. There is no doubt that he independently solved the problem of the cube and the cosa, but the glory is commonly divided between del Ferro, who had first cracked one type of cubic, and Cardano, who mastered cubics in all their generality and was godfather to the solution of the quartic.

Chapter 5

RELIEF FOR THE IMAGINATION

§5.1   THERE WERE TWO GREAT ADVANCES in algebra across the early modern period in Europe, by which I mean the two centuries from the fall of Constantinople (1453) to the Peace of Westphalia (1648). They were (1) the solution of the general cubic and quartic equations and (2) the invention of modern literal symbolism—the systematic use of letters to stand for numbers.

The first of those advances was accomplished by northern Italian mathematicians from about 1520 to 1540, the period of most concentrated creativity probably being the joint deliberations of Cardano and Ferrari in 1539–1540. That is the story I told in Chapter 4.

The second was largely the work of two Frenchmen: François Viète[50] (1540–1603) and René Descartes (1596–1650). It proceeded in parallel with another development: the slow discovery of complex numbers, and their gradual acceptance as part of the standard mathematical toolbox. This latter development was more properly arithmetic (concerned with numbers) than algebraic (concerned with polynomials and equations). As we have seen, though, it drew its inspiration from algebra. If you graph an "irreducible" cubic polynomial (see Figure CQ-6), it plainly has three real zeros; yet if you apply the algebraic formulas to solve the corresponding equation and disallow complex numbers, there are no real solutions at all!

There are other reasons to offer complex numbers a guest ticket to the history-of-algebra party. They are, for example, the first hint of the key algebraic concept of *linear independence,* which I shall discuss later and which led to the theories of vectors and tensors, making modern physics possible. If you add 3 to 5, you get 8: the three-ness of the 3 and the five-ness of the 5 have merged and been lost in the eight-ness of the 8, like two droplets of water coalescing. If, however, you add 3 to 5*i,* you get the complex number 3 + 5*i,* a droplet of water and a droplet of oil—linear independence.

I shall therefore say what is necessary and interesting to say about the discovery of complex numbers. The first mathematician to take these strange creatures even half-seriously was Cardano, as we have seen. The first to tackle them with any confidence was Rafael Bombelli.

§5.2   Bombelli was from Bologna, where Scipione del Ferro had taught. He was born in 1526, the year del Ferro died. He was therefore a clear generation younger than Cardano. As is often the case, what was a struggle for one generation to grasp came much more easily to the following generation. Bombelli would have been 19 when *Ars magna* appeared—just the right age to be receptive to its influence.

Bombelli was a civil engineer by trade. His first big commission was land reclamation work, draining some marshes near Perugia in central Italy. This task took from 1549 to 1560. It was a great success and made Bombelli's name in his profession.

Bombelli admired *Ars magna* but felt that Cardano's explanations were not clear enough. At some point in his 20s, he conceived the ambition to write an algebra book of his own, one that would enable a complete beginner to master the subject. The book, titled *l'Algebra,* was published in 1572, a few months before Bombelli's death, so presumably he was working on it for a quarter of a century, from his early 20s to his mid-40s. The work must have gone through many changes and revisions, but one that we know of is particularly noteworthy.

Around 1560, Bombelli was in Rome. There he met and talked math with Antonio Maria Pazzi, who taught the subject at the university in that city. Pazzi mentioned that he had found in the Vatican library a manuscript on arithmetic and algebra by "a certain Diophantus," a Greek author of the ancient world. The two men examined the text and decided to make a translation of it. The translation was never finished, but there is no doubt Bombelli got much inspiration from studying Diophantus. He included 143 of Diophantus's problems in *l'Algebra*, and it was through his book that Diophantus's work first became known to European mathematicians of the time.

Recall that Diophantus, though he had no conception of negative numbers as mathematical objects in their own right and would not accept them as solutions to problems, nonetheless allowed them a shadowy existence in his intermediate calculations and formulated the rule of signs for this purpose. Cardano seems to have regarded complex numbers in a similar fashion, as having no meaning in themselves but being useful devices for getting from a real problem to a real solution.

Bombelli's approach to negative and complex numbers was more mature. Negative numbers he took at face value, restating the rule of signs more clearly than had Diophantus:

> *più via più fa più*
> *meno via più fa meno*
> *più via meno fa meno*
> *meno via meno fa più*

Here *più* means "positive," *meno* "negative," *via* "times," and *fa* "makes."

In *l'Algebra*, Bombelli takes on the irreducible cubic, finding a solution of the equation $x^3 = 15x + 4$. Using Cardano's method, he gets

$$x = \sqrt[3]{2 + \sqrt{-121}} + \sqrt[3]{2 - \sqrt{-121}}$$

By some ingenious arithmetic, he works out the cube roots to be $2 + \sqrt{-1}$ and $2 - \sqrt{-1}$, respectively. Adding these, he gets the solution $x = 4$. (The other solutions, which he does not get, are $-2 - \sqrt{3}$ and $-2 + \sqrt{3}$.)

The complex numbers here are like Diophantus's negative numbers—a sort of internal trick for getting from a "real" problem to a "real" solution. They are, as it were, catalytic. "Sophistic" is what Bombelli actually called them. He accepted them as legitimate working devices, though, and even gave a sort of "rule of signs" for multiplying them:

> *più di meno via più di meno fa meno*
> *più di meno via meno di meno fa più*
> *meno di meno via più di meno fa più*
> *meno di meni via meno di meno fa meno*

Here *più di meno*, "positive from negative," means $+\sqrt{-N}$, while *meno di meno*, "negative from negative," means $-\sqrt{-N}$, N being some positive number, *via* and *fa* as before. So the third line of the jingle means: If you multiply $-\sqrt{-N}$ by $+\sqrt{-N}$, you will get a positive result. This is quite true: The result will be N. The ordinary rule of signs (– times +) gives us a negative; squaring the square root gives us $-N$, and the negative of $-N$ is N.

Bombelli's *l'Algebra* is a great step forward in mathematical understanding, but he was still held back by lack of a good symbolism. For the formula

$$\sqrt[3]{2 + \sqrt{-3}} \times \sqrt[3]{2 - \sqrt{-3}} \, ,$$

he writes

> *Moltiplichisi*, R.c. ⌊2 *più di meno* R.q.3⌋ per
> R.c. ⌊2 *meno di meno* R.q.3⌋

Here "R.q." means a square root, "R.c." a cube root. Note the use of brackets. As symbolism goes, this is an improvement on Cardano's generation but not by much.

§5.3   The 16th century was not a happy time to be a French citizen. Much of the reign of Francis I (1515–1547)—and much of the national wealth, too—was consumed by wars against the Emperor Charles V. No sooner were the combatants thoroughly exhausted ('Treaty of Cateau-Cambrésis, 1559) than French Catholics and Protestants—the latter commonly called Huguenots[51]—set to massacring one another.

They continued to do so until the Edict of Nantes (1598) put an end, or at any rate an 87-year pause, to it all. The previous 36 years had seen eight civil wars among the French and a change of dynasty to boot (Valois to Bourbon, 1589). These wars were not purely religious. Elements of regional sentiment, social class, and international politics played their parts. Philip II of Spain, one of the greatest troublemakers of all time, did his best to keep things boiling.[52] So far as class was concerned, the Huguenots were strong among the urban middle classes, but much of the nobility—perhaps a half—were Protestant, too. Peasants, by contrast, remained overwhelmingly Catholic in most regions.

François Viète was born in 1540 into a Huguenot family. His father was a lawyer. He graduated with a law degree from the University of Poitiers in 1560. The French wars of religion began less than two years later, with a massacre of Huguenots at Vassy in the Champagne region.

Viète's subsequent career was shaped by the wars. He gave up lawyering to become tutor to an aristocratic family. Then in 1570, he moved to Paris, apparently in the hope of government employment. The young Charles IX was king at this time, but his mother Catherine de' Medici (who was also the mother-in-law of Philip II of Spain) was the real power center. Her policy of playing off Huguenots against

Catholics in order to keep the throne strong and independent of all factions determined the course of French history through the 1560s, 1570s, and 1580s, and often produced paradoxical results. Thus Viète was in Paris when Charles authorized the general massacre of Huguenots on St. Bartholomew's Eve (August 23, 1572); yet the following year, Viète, a Huguenot, was appointed by the king to a government position in Brittany.

Charles died in 1574, to be succeeded by Henry III, Catherine's third son. Viète returned to Paris six years later to take up a position as adviser to this king. Catherine's youngest son died in 1584, however, leaving the Valois line without an heir. Henry III, though he was married and only 33 years old, was flamboyantly gay, wont to show up at court functions in drag. It was thought unlikely that he would produce a son. That left his distant relative Henry of Navarre, of the Bourbon family, as lawful heir to the throne. That Henry, however, was a Protestant, a fact that alarmed Catholics both inside and outside France. Infighting at the court became very intense. Viète was forced out and obliged to take a five-year sabbatical in his home district, at the little town of Beauvoir-sur-Mer on the Bay of Bourgneuf. This period, 1584–1589, was Viète's most mathematically creative—unusual as mathematical creativity goes, for he was in his late 40s. The court politics of France at this point were so convoluted that it is hard for the historian of mathematics to know whom to thank.

Just four months after Viète's return to court, Henry III was assassinated, stabbed while sitting on his commode. Henry of Navarre became Henry IV, first king of the Bourbon dynasty. The fact of the new king's being a Protestant suited Viète, who happily joined his entourage. The Catholics, however, were not about to allow Henry IV's accession without a fight, even though they could not agree on a rival candidate for the throne. Philip of Spain favored his own daughter and intrigued with factions at the French court on her behalf. These intrigues relied on letters written in a code. Finding he had a mathematician at hand, Henry set Viète the task of cracking the Spanish code. Viète, after some months of effort, finally did so. When it dawned on Philip that his unbreakable code had been broken, he complained to the Pope that Henry was using witchcraft.

§5.4   Viète continued to serve Henry IV until he was dismissed from the court in December of 1602. He then returned to his hometown and died a year later.

Next to Viète's cryptographic triumph, the mathematical high point of his royal service came in 1593. In that year the Flemish mathematician Adriaan van Roomen published a book titled *Ideae mathematicae*, which included a survey of all the prominent mathematicians of the day. The Dutch ambassador to Henry IV's court pointed out to Henry that not a single French person was listed. To drive the point home, he showed the king a problem in Roomen's book, one for the solution of which the author was offering a prize. The problem was to find numbers x satisfying an equation of the 45th degree, beginning $x^{45} - 45x^{43} + 945x^{41} - 12300x^{39}$. . . . Surely, sneered the diplomat (who seems not to have been very diplomatic), no French mathematician could solve this problem. Henry sent for Viète, who found a solution on the spot and came up with 22 more the following day.

Viète knew, of course, that Roomen had not just given any old random equation. It had to be one that Roomen himself knew how to solve. A man of his time, Viète also had his head full of trigonometry, a great mathematical growth point just then.[53] His first two books had been collections of trigonometric tables. Trigonometry—the study of numerical relations between the arc lengths and chord lengths of a circle—is full of long formulas involving sines, cosines, and their powers. Some speedy mental arithmetic on the first few coefficients in the equation would have told Viète that he was looking at just such a formula: the polynomial for $2\sin 45\alpha$ in terms of $x = 2\sin\alpha$. Trigonometry then gave him the solutions. (At least it gave him the 23 positive solutions. There are also 22 negative ones, which Viète ignored, apparently considering them meaningless.)

§5.5   The fruit of those five years of seaside exile when Viète was in his 40s was a book titled *In artem analyticem isagoge* ("Introduction to the Analytic Art"). The *Isagoge* represents a great step forward in algebra and a small step backward. The forward step was the first systematic use of letters to represent numbers. The germ of this idea goes back to Diophantus, but Viète was the first to deploy it effectively, making a range of letters available for many different quantities. Here is the beginning of modern literal symbolism.

Viète's literal symbolism was not restricted to the unknown quantity, as all previous such schemes had been. He divided quantities into two classes: unknown quantities, or "things sought" (*quaesita*), and known ones, or "things given" (*data*). The unknowns he denoted by uppercase vowels *A, E, I, O, U,* and *Y*. The "things given" he denoted by uppercase consonants: *B, C, D,* . . . . Here, for example, is the equation $bx^2 + dx = z$ in Viète's symbolism:

>   *B* in *A* Quadratum, plus *D* plano in *A*, aequari *Z* solido.

His *A*, the unknown, is our *x*. The other symbols are all *data*.

That "plano" and "solido" show the backward step I mentioned. Viète was strongly influenced by the geometry of the ancients and wanted to base his algebra rigorously on geometrical concepts. This, as he saw it, obliged him to follow a *law of homogeneity*, obliging every term in an equation to have the same dimension. Unless otherwise specified, every symbol stands for a line segment of the appropriate length. In the equation given above, *b* and *x* (Viète's *B* and *A*) therefore have one dimension each. It follows that bx^2 has three dimensions. Therefore *dx* must also have three dimensions, and so must *z*. Since *x* is a one-dimensional line segment, *d* must be two-dimensional—hence, "*D* plano." Similarly, *z* must be three-dimensional: "*Z* solido."

You can see Viète's point, but this law of homogeneity cramps his style and makes some of his algebra difficult to follow. It also seems a little odd that a man so deft with polynomials of the 45th degree should have rooted himself so firmly in classical geometry and its mere three dimensions.

§5.6   Viète's treatment of equations was in some ways less "modern" than Bombelli's. He was, as I have said, averse to negative numbers, which he did not admit as solutions. His attitude to complex numbers was even more retrograde. He did deal with cubic equations, but in a book about geometry, where he offers a trigonometric solution based on the formula for $\sin 3\alpha$ in terms of $\sin \alpha$.

In one respect, though, Viète was a pioneer in the study of equations, and lit a candle which, 200 years later, flared into a mighty beacon. This particular discovery was not published in his lifetime. Twelve years after Viète's death, his Scottish friend Alexander Anderson published two of his papers on the theory of equations. In the second paper, titled *De equationem emendatione* ("On the Perfecting of Equations"), Viète opened up the line of inquiry that led to the study of the symmetries of an equation's solutions, and therefrom to Galois theory, the theory of groups, and all of modern algebra.

Consider the quadratic equation $x^2 + px + q = 0$. Suppose the two solutions of this equation—the actual numbers that make it true—are α and β. If x is α or x is β, and never otherwise, the following thing must be true:

$$(x - \alpha)(x - \beta) = 0$$

Since α and β, and no other values of x at all, make this equation true, it must be just a rewritten form of the equation we started with. Now, if you multiply out those parentheses in the usual way, this rewritten equation amounts to

$$x^2 - (\alpha + \beta)x + \alpha\beta = 0$$

Comparing this equation with the original one, it must be the case that

$$\alpha + \beta = -p$$
$$\alpha\beta = q$$

Here we have relationships between the *solutions* of the equation and the *coefficients*. You can do a similar thing for the cubic equation $x^3 + px^2 + qx + r = 0$. If the solutions of this equation are α, β, and γ, then

$$\alpha + \beta + \gamma = -p$$
$$\beta\gamma + \gamma\alpha + \alpha\beta = q$$
$$\alpha\beta\gamma = -r$$

It works for the quartic $x^4 + px^3 + qx^2 + rx + s = 0$, too:

$$\alpha + \beta + \gamma + \delta = -p$$
$$\alpha\beta + \beta\gamma + \gamma\delta + \alpha\gamma + \beta\delta + \alpha\delta = q$$
$$\beta\gamma\delta + \gamma\delta\alpha + \delta\alpha\beta + \alpha\beta\gamma = -r$$
$$\alpha\beta\gamma\delta = s$$

And for the quintic $x^5 + px^4 + qx^3 + rx^2 + sx + t = 0$:

$$\alpha + \beta + \gamma + \delta + \varepsilon = -p$$
$$\alpha\beta + \beta\gamma + \gamma\delta + \delta\varepsilon + \varepsilon\alpha + \alpha\gamma + \beta\delta + \gamma\varepsilon + \delta\alpha + \varepsilon\beta = q$$
$$\gamma\delta\varepsilon + \alpha\delta\varepsilon + \alpha\beta\varepsilon + \alpha\beta\gamma + \beta\gamma\delta + \beta\delta\varepsilon + \alpha\gamma\varepsilon + \alpha\beta\delta + \beta\gamma\varepsilon + \alpha\gamma\delta = -r$$
$$\beta\gamma\delta\varepsilon + \gamma\delta\varepsilon\alpha + \delta\varepsilon\alpha\beta + \varepsilon\alpha\beta\gamma + \alpha\beta\gamma\delta = s$$
$$\alpha\beta\gamma\delta\varepsilon = -t$$

The correct way to read those lines is:

All possible singletons added together $= -p$
All possible products of pairs added together $= q$
All possible products of triplets added together $= -r$
etc.

These things were first written down by Viète, for these first five degrees of equations in a single unknown. A French mathematician of the following generation, Albert Girard, generalized them to an equation of any degree in his book *New Discoveries in Algebra*, pub-

lished in 1629, 14 years after Anderson's publication of Viète's paper.
Sir Isaac Newton picked them up, and . . . but I am getting ahead of
my story.

§5.7 René Descartes needs, as emcees say, no introduction. Soldier
and courtier (he survived the first but not the second), philosopher
and mathematician, a French subject under the first three Bourbon
kings, his adult life was spent in the time of the Thirty Years War, the
English Civil War, and the Pilgrim fathers; the time of Cardinal
Richelieu of France and King Gustavus Adolphus of Sweden, of
Milton and Galileo. He is one of the national heroes of France, though
he preferred to live in the Netherlands. His birthplace, at that time
named La Haye, was renamed Descartes in his honor after the French
Revolution. (It is 30 miles northeast of Poitiers.) He started out in the
world with a law degree from Poitiers University, just as Viète had 56
years earlier.

Descartes is popularly known for two things: for having written
Cogito ergo sum ("I think, therefore I am") and for the system named
after him that identifies all the points of a plane by numbers. In Cartesian—Descartes' Latin name was Cartesius—geometry, the numbers that identify a point are the perpendicular distances of that point
from two fixed lines drawn across the plane at right angles to each
other. The west-east distance is conventionally called x, the south-
north distance y. These are the Cartesian coordinates of a point (see
Figure 5-1).

In fact, although Descartes did write *Cogito ergo sum*, he did not
precisely invent the Cartesian system of coordinates. The main idea
of it is there in his work *La géométrie* (1637), but the baselines he uses
are not at right angles to each other.

The ideas contained in *La géométrie* did, however, suffice to revo-
lutionize both algebra and geometry—to algebraize geometry, in fact.
Recall Viète's law of homogeneity from §5.5, which rested on the idea
that numbers are, fundamentally, the thought-shadows of geometri-

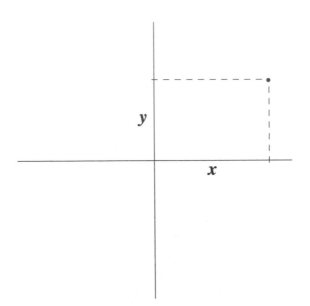

FIGURE 5-1 Cartesian coordinates.

cal objects. Even when presented with a 45th power, Viète's mind turned at once to a geometrical notion—a circular arc divided into 45 equal parts. Descartes stood this on its head. Geometrical objects, he showed, might just be convenient representations of numbers. The product of two line segment lengths need not be thought of as the area of some rectangle; it can be represented by another line segment—Descartes gave a convincing example.

This was not an especially original thought in itself, but by proceeding to build his entire scheme of geometry on it, Descartes chopped through the last hawsers connecting algebra to classical geometry, allowing his new "analytic geometry" to soar up into the heavens. It was further enabled to do so by Descartes' adoption of a clearer and more manageable system of algebraic notation. He took up the plus and minus signs from the German Cossists of the previous century, and also their square root sign (to which he added the overbar, turning $\sqrt{\ }$ into $\sqrt{\ }$). He used superscripts for exponentiation, though not for squares, which he mostly wrote as *aa*, instead of a^2—a practice some mathematicians continued into the late 19th century.

Perhaps most important, Descartes gave us the modern system of literal symbolism, in which lowercase letters from the beginning of the alphabet are used to stand for numbers given, the *data*, and letters from the end of the alphabet for numbers sought, the *quaesita*. Art Johnson, in his book *Classic Math*, has a story about this.

> The predominant use of the letter x to represent an unknown value came about in an interesting way. During the printing of *La géométrie*... the printer reached a dilemma. While the text was being typeset, the printer began to run short of the last letters of the alphabet. He asked Descartes if it mattered whether x, y, or z was used in each of the book's many equations. Descartes replied that it made no difference which of the three letters was used to designate an unknown quantity. The printer selected x for most of the unknowns, since the letters y and z are used in the French language more frequently than is x.

Reading *La géométrie*, in fact, you feel that you are looking at a modern mathematical text. It is the earliest book for which this is true, I think. The only real oddity is the absence of our modern equals sign: Descartes used a little symbol like an infinity sign with the left end cut off.

The introduction of a good workable literal symbolism was a great advance in mathematics. It was not Descartes' alone—we have seen how Viète began the systematic use of letters for numbers, and in the case of unknowns the original inspiration goes back to Diophantus.

It would be unjust not to mention the Englishman John Harriot here, too. Harriot, who lived from 1560 to 1621, was of the generation between Viète and Descartes. He spent many years in the service of Sir Walter Raleigh, traveling on at least one of Raleigh's expeditions to Virginia. He was a keen mathematician and, likely under the inspiration of Viète, used letters of the alphabet for both *data* and *quaesita*. Fluent in the theory of equations, Harriot considered both negative and complex solutions. Unfortunately, none of this came to light until some years after his death,[54] as he published no math while alive. Historians of math like to debate how much Descartes borrowed from Harriot—*La géométrie* appeared six years after Harriot's algebraic work was published (in a clumsily edited form). So far as I know, however, no one has been able to reach a firm conclusion on the matter of Descartes' debt to Harriot.

It was, at any rate, Descartes who first made widely known and available a system of literal symbolism robust enough to need no substantial changes over the next four centuries. Not only was this a boon to mathematicians, it inspired Leibniz's dream of a symbolism for all of human thought, so that all arguments about truth or falsehood could be resolved by calculation. Such a system would, said Leibniz, "relieve the imagination." When we compare Descartes' mathematical demonstrations with the wordy expositions of earlier algebraists, we see that a good literal symbolism really does relieve the imagination, reducing complex high-level thought processes to some easily mastered manipulations of symbols.

In 1649, Queen Christina of Sweden, Gustavus Adolphus's daughter, persuaded Descartes to teach her philosophy. She sent a ship to fetch him, and Descartes took up residence with the French ambassador in Stockholm. Unfortunately, the queen was an early riser, while Descartes had been accustomed since childhood to lie in bed until 11 a.m. Trudging across the windy palace squares at 5 a.m. through the bitter Swedish winter of 1649–1650, Descartes caught pneumonia and died on February 11 that latter year. Isaac Newton was just seven years old.

Part 2

UNIVERSAL ARITHMETIC

Chapter 6

THE LION'S CLAW

§6.1   THE BRITISH ISLES PRODUCED some fine mathematicians during the period from the late 16th century to the early 18th century, despite being busy with a civil war (1642–1651), a military dictatorship (1651–1660), a constitutional revolution (1688), and two changes of dynasty (Tudor to Stuart in 1603, and Stuart to Hanoverian in 1714).

I have already mentioned Thomas Harriot, whose sophisticated literal symbolism went largely unnoticed (except perhaps by Descartes). The Scotsman John Napier, though not significant as an algebraist, had discovered logarithms and presented them to the world in 1614. He also popularized the decimal point. William Oughtred, an English country parson, wrote on algebra and trigonometry and gave us the × sign for multiplication. John Wallis was the first to take up Descartes' techniques and notations for analytical geometry (though he was a champion of Harriot, now long dead, and a vigorous proponent of the opinion that Descartes had gotten his notation from Harriot).

All these figures were, however, mere prologues to the arrival of Isaac Newton. This tremendous genius, by common agreement the greatest name in the history of science, was born on Christmas Day 1642,[55] to the widow of a prosperous farmer in Lincolnshire. The subsequent course of his life and the character of the man have been much written about. Here are some words I myself wrote on those topics.

> The story of Newton's life . . . is not enthralling. He never traveled outside eastern England. He took no part in business or in war. In spite of having lived through some of the greatest events in English constitutional history, he seems to have had no interest in public affairs. His brief tenure as a Member of Parliament for Cambridge University made not a ripple on the political scene. Newton had no intimate connections with other human beings. On his own testimony, which there is no strong reason to doubt, he died a virgin. He was similarly indifferent to friendship, and published only with reluctance, and then often anonymously, for fear that: "[P]ublic esteem, were I able to acquire and maintain it . . . would perhaps increase my acquaintance, the thing which I chiefly study to decline." His relationships with his peers, when not tepidly absent-minded, were dominated by petty squabbles, which he conducted with an irritated punctiliousness that never quite rose to the level of an interesting vehemence. "A cold fish," as the English say.[56]

I cannot resist at this point telling my favorite Newton story, though I think it is quite well known. In 1696 the Swiss mathematician Johann Bernoulli posed two difficult problems to the mathematicians of Europe. Newton solved the problems the day he was shown them and passed on his solutions to the president of the Royal Society in London, who sent them to Bernoulli without telling him who had supplied them. As soon as he read the anonymous solutions, Bernoulli knew them to be Newton's—"*tanquam ex ungue leonem*," he said ("as by [his] claw [we know] the lion").

That mighty claw scratched one great mark across the history of algebra.

§6.2 Newton[57] is famous for his contributions to science and for having invented calculus, but he is not well known as an algebraist. In fact, he had lectured on algebra at Cambridge University from 1673 to 1683 and deposited his lecture notes in the university library. Many years afterward, when he had left academic life and was established as master of the Royal Mint, his Cambridge successor, William Whiston, published the lectures as a book with the title *Arithmetica universalis* (Universal Arithmetic). Newton gave his permission for this publication, but only reluctantly, and he seemed never to have liked the book. He refused to have his name appear as author and even contemplated buying up the whole edition himself so that he could destroy it. Nor did Newton's name appear on the English version (*Universal Arithmetic*), published in 1720, nor on the second Latin edition in 1722.[58]

What most excites the interest of a historian of algebra, though, is not the *Universal Arithmetic* but some jottings a much younger Newton put down in 1665 or 1666 and that can be found in the first volume of his *Collected Mathematical Works*. They are in English, not Latin, and commence with the words:

> Every Equation as $x^8 + px^7 + qx^6 + rx^5 + sx^4 + tx^3 + vxx + yx + z = 0$. hath so many roots as dimensions, of wch y^{e} summ is $-p$, the summ of the rectangles of each two $+q$, of each three $-r$, of each foure $+s$ . . . .

These notes do not state any theorem. They suggest one, though; and the theorem is such a striking one that on the strength of the suggestion, mathematicians (as well as, in fact, the editor of the *Collected Works*) speak of Newton's theorem.

Before presenting the theorem, I need to explain the concept of a *symmetric* polynomial. To keep things manageable, I'll consider just three unknowns, calling them α, β, and γ. Here are some symmetric polynomials in these three unknowns:

$$\alpha\beta + \beta\gamma + \gamma\alpha$$
$$\alpha^2\beta\gamma + \alpha\beta^2\gamma + \alpha\beta\gamma^2$$
$$5\alpha^3 + 5\beta^3 + 5\gamma^3 - 15\alpha\beta\gamma$$

Here are some polynomials in α, β, and γ that are *not* symmetric:

$$\alpha\beta + 2\beta\gamma + 3\gamma\alpha$$
$$\alpha\beta^2 - \alpha^2\beta + \beta\gamma^2 - \beta^2\gamma + \gamma\alpha^2 - \gamma^2\alpha$$
$$\alpha^3 - \beta^3 - \gamma^3 + 2\alpha\beta\gamma$$

What distinguishes the first group from the second? Well, just by eyeball inspection, something like this: In the first group, everything that happens to α happens likewise to β and γ, everything that happens to β happens likewise to γ and α, and everything that happens to γ happens likewise to α and β. Things—addition, multiplication, combination—are happening to all three unknowns in a very even-handed way. This is not the case in the second group.

That is pretty much it, but the condition of being a symmetric polynomial can be described with more mathematical precision: *If you permute α, β, and γ in any way at all, you end up with the same expression.*

There are in fact five ways to permute α, β, and γ:

- Switch β and γ, leaving α alone.
- Switch γ and α, leaving β alone.
- Switch α and β, leaving γ alone.
- Replace α by β, β by γ, and γ by α.
- Replace α by γ, β by α, and γ by β.

(Note: A mathematician would try to persuade you that there are six permutations, the sixth being the "identity permutation," where you don't do anything at all. I shall adopt this point of view myself in the next chapter.)

If you were to do any of those things to any one of that first group of polynomials, you would end up with just the polynomial you started with, though perhaps in need of rewriting. If you do the last permutation to $\alpha\beta + \beta\gamma + \gamma\alpha$, for example, you get $\gamma\alpha + \alpha\beta + \beta\gamma$, which is the same thing, but written differently.

Another way of looking at this, and a useful (though not perfectly infallible!) way to check for symmetry when the polynomials are big and unwieldy, is to assign random numbers to α, β, and γ. Then the polynomial works out to a single numerical value. If this value is the same when you assign the same numbers to α, β, and γ in all possible different orders, it is symmetrical. If I assign the arbitrary numbers 0.55034, 0.81217, and 0.16110 to α, β, and γ in all six possible ways and work out the corresponding six values of $\alpha\beta^2 - \alpha^2\beta + \beta\gamma^2 - \beta^2\gamma + \gamma\alpha^2 - \gamma^2\alpha$, I get 0.0663536 and −0.0663536 three times each. *Not* a symmetric polynomial: Permuting the unknowns gives two different values of the polynomial. (An interesting thing in itself—why two?—which I shall say more about later.)

All these ideas can be extended to any number of unknowns and to expressions of any level of complexity. Here is a symmetric polynomial of the 11th degree in two unknowns:

$$\alpha^8\beta^3 + \alpha^3\beta^8 - 12\alpha - 12\beta$$

Here is a symmetric polynomial of the second degree in 11 unknowns:

$$\alpha^2 + \beta^2 + \gamma^2 + \delta^2 + \varepsilon^2 + \zeta^2 + \eta^2 + \theta^2 + \iota^2 + \kappa^2 + \lambda^2$$

Now, not all symmetric polynomials are equally important. There is a subclass called the *elementary* symmetric polynomials. For three unknowns the elementary symmetric polynomials are

Degree 1: $\alpha + \beta + \gamma$
Degree 2: $\beta\gamma + \gamma\alpha + \alpha\beta$
Degree 3: $\alpha\beta\gamma$

Of the examples I gave of symmetric polynomials above, the first is elementary, the other two are not.

You can think of the elementary symmetric polynomials in any number of unknowns as

Degree 1: All the unknowns, added together ("all singletons").
Degree 2: All possible pairs, added together ("all pairs").

Degree 3: All possible triplets, added together ("all triplets").
Etc.

If you are working with n unknowns, the list runs out after n lines because you can't make an $(n + 1)$-tuplet out of n unknowns. Now I can show you Newton's theorem.

Newton's Theorem

Any symmetric polynomial in n unknowns
can be written in terms of
the *elementary* symmetric polynomials in n unknowns.

So although the other two examples of symmetric polynomials that I gave are not elementary, they can be written in terms of the three elementary symmetric polynomials I just showed. The second one is easy:

$$\alpha^2\beta\gamma + \alpha\beta^2\gamma + \alpha\beta\gamma^2 = \alpha\beta\gamma(\alpha + \beta + \gamma)$$

The third is a little trickier, but you can easily confirm that

$$5\alpha^3 + 5\beta^3 + 5\gamma^3 - 15\alpha\beta\gamma$$
$$= 5(\alpha + \beta + \gamma)^3 - 15(\alpha + \beta + \gamma)(\beta\gamma + \gamma\alpha + \alpha\beta)$$

By convention, and leaving it understood that we are dealing with some fixed number of unknowns (in this case three), the elementary symmetric polynomials are denoted by lowercase Greek sigmas, with a subscript to indicate degree. In this case, with three unknowns, the degree-1, degree-2, and degree-3 elementaries are called σ_1, σ_2, and σ_3. So I could write that last identity as

$$5\alpha^3 + 5\beta^3 + 5\gamma^3 - 15\alpha\beta\gamma = 5\sigma_1^3 - 15\sigma_1\sigma_2$$

There is Newton's theorem: Any symmetric polynomial in any number of unknowns can be written in terms of the sigmas, the *elementary* symmetric polynomials.

§6.3   What has all this got to do with solving equations? Why, just look back at those polynomials in α, β, γ, and so on, in §5.6, the ones Viète tinkered with. They are the elementary symmetric polynomials! For the general quintic equation $x^5 + px^4 + qx^3 + rx^2 + sx + t = 0$, if the solutions are $\alpha, \beta, \gamma, \delta$, and ε, then $\sigma_1 = -p, \sigma_2 = q, \sigma_3 = -r, \sigma_4 = s$, and $\sigma_5 = -t$, where the sigmas are the elementary symmetric polynomials in five unknowns, which I actually wrote out in §5.6. A similar thing is true for the general equation of any degree in x.

Those jottings of Newton's, the ones that lead us to Newton's theorem, were, as I mentioned, done in 1665 or 1666, very early in Newton's mathematical career. This was the time when, at age 21, just after he had obtained his bachelor's degree, Newton had to go back to his mother's house in the countryside because an outbreak of plague had forced the University of Cambridge to close. Two years later the university reopened, and Newton went back for his college fellowship and master's degree. During those two years in the countryside, Newton had worked out all the fundamental ideas that underlay his discoveries in math and science. It is not quite the case, as folklore has it, that mathematicians never do any original work after age 30, but it is generally true that their style of thinking, and the topics that attract their keenest interest, can be found in their early writings.

Newton actually had a particular problem in mind when making those early jottings, the problem of determining when two cubic equations have a solution in common. However, this work on

(1) symmetric polynomials in general, and

(2) the relationships between the *coefficients* of an equation, and symmetric polynomials in the *solutions* of that equation

was crucial to further development of the theory of equations, and all that flowed from it, both within that theory and then beyond it into whole new regions of algebra. Symmetry . . . polynomials in the solutions as expressions in the coefficients . . . these were the keys to solving the great outstanding problem in the theory of polynomial equations at this point in the later 17th century, 120 years since the cracking of the cubic and the quartic: to find an algebraic solution for the general quintic.

§6.4   Speaking very generally, the 18th century was a slow time for algebra, at any rate by comparison with the 17th and 19th centuries. The discovery of calculus by Newton and Leibniz in the 1660s and 1670s opened up vast new mathematical territories for exploration, none of them algebraic in the sense I am using in this book. The area of math we now call "analysis"—the study of limits, infinite sequences and series, functions, derivatives, and integrals—was then new and sexy, and mathematicians took to it with enthusiasm.

There was a more general mathematical awakening, too. The modern literal symbolism developed for algebra by Viète and Descartes made all mathematical work easier by "relieving the imagination." Furthermore, the growing acceptance of complex numbers stretched the imaginative boundaries of math. De Moivre's theorem, which first appeared in finished form in 1722, may be taken as representative of early 18th-century pure mathematics. Stating that

$$(\cos\theta + i\sin\theta)^n = \cos n\theta + i\sin n\theta,$$

the theorem threw a bridge between trigonometry and analysis and helped make complex numbers indispensable to the latter.

That is only to speak of pure mathematics. With the rise of science, the first stirrings of the Industrial Revolution, and the settling down of the modern European nation-system after the wars of religion, mathematicians were increasingly in demand by princes and generals. Euler designed the plumbing for Frederick the Great's palace at Sans Souci; Fourier was a scientific adviser on Napoleon's expedition to Egypt.

D'Alembert's pioneering work on differential equations in the middle of the century was characteristic, and Laplace's equation $\nabla^2 \phi = 0$, which describes numerous physical systems where a quantity (density, temperature, electric potential) is distributed smoothly but unequally across an area or a volume, can be taken as representative of applied math at the century's end.

Algebra was something of a bystander to all these glamorous developments. The general cubic and quartic equations had been cracked, but no one had much of a clue about how to proceed further in that direction. Viète, Newton, and one or two others among the most imaginative mathematicians had noticed the odd symmetries of the solutions of polynomial equations but had no idea how to make any mathematical profit from these observations.

There was, however, one other problem that mathematicians struggled with all through the 18th century and that I ought to cover here. This was the problem of finding a proof for the so-called fundamental theorem of algebra, hereinafter the FTA. I write "so-called" because the theorem always *is* so called, yet its status as implied by that name is considerably disputed. There are even mathematicians who will tell you, in the spirit of Voltaire's well-known quip about the Holy Roman Empire, that the FTA is neither fundamental, nor a theorem, nor properly within the scope of algebra. I hope to clarify all that in just a moment.

The FTA can be stated very simply, if a little roughly, in the context of polynomial equations, as: *Every equation has a solution.* To be more precise:

The Fundamental Theorem of Algebra

The polynomial equation $x^n + px^{n-1} + qx^{n-2} + \ldots = 0$ in
a single unknown x, the polynomial's coefficients
$p, q, \ldots$ being complex numbers, and n greater than zero,
is satisfied by some complex number.

Ordinary real numbers are to be understood here as just particular cases of complex numbers, the real number a as the complex number $a + 0i$. So equations with real-number coefficients, like all the ones I have displayed so far, come under the scope of the FTA. Every such equation has a solution, though the solution may be a complex number, as in the case $x^2 + 1 = 0$, satisfied by the complex number i (and also by the complex number $-i$).

The FTA was first stated by Descartes in *La géométrie* (1637), though in a tentative form, as he was not at ease with complex numbers. All the great 18th-century mathematicians had a go at trying to prove it. Leibniz actually thought he had *dis*proved it in 1702, but there was an error in his reasoning, pointed out by Euler 40 years later. The mighty Gauss made it the subject of his doctoral dissertation in 1799. Not until 1816 was a completely watertight proof given, though—also by Gauss.

To clarify the mathematical status of the FTA, you really need to study a proof. The proof is not difficult, once you have made friends with the complex plane (see Figure NP-4) and can be found in any good textbook of higher algebra.[59] What follows is the merest outline.

§6.5 *Proof of the Fundamental Theorem of Algebra*

It is the case with complex numbers, as with real numbers, that higher powers easily swamp lower ones, a thing I mentioned in §CQ.3. Cubes get seriously big much faster than squares, and fourth powers much faster than cubes, and so on. (Note: The word "big," when applied to complex numbers, means "far from the origin," or equivalently "having a large modulus.") For big values of x, therefore, the polynomial in that box above looks pretty much like x^n with some small adjustments caused by the other terms.

If x is zero, on the other hand, every term in the polynomial is equal to zero, except for the last, "constant," term. So for tiny values of x, the polynomial just looks like that last constant term. (The constant term in $x^2 + 7x - 12$, for instance, would be -12.)

If you change x smoothly and evenly, then x^2, x^3, x^4, and all higher powers will also change smoothly and evenly, though at different speeds. They will not suddenly "jump" from one value to another.

Given those three facts, consider *all* the complex numbers x with some given large modulus M. These numbers, if you mark them in the complex plane, form the circumference of a perfect circle of radius M. The corresponding values of the polynomial form, but only approximately, the circumference of a much bigger circle, one with radius M^n. (If a complex number has modulus M, its square has modulus M^2 and so on. This is easy to prove.) That's because x^n has swamped all the lower terms of the polynomial.

Gradually, smoothly, shrink M down to zero. Our perfect circle— all the complex numbers with modulus M—shrinks down to the origin. The corresponding values of the polynomial shrink down correspondingly, like a loop of rope tightening, from a vast near-circle centered on the origin to the single complex number that is the constant term in the polynomial. And in shrinking down like this, the tightening polynomial loop must at some point cross the origin. How else could all its points end up at that one complex number?

Which proves the theorem! The points of that dwindling loop are values of the polynomial, for some complex numbers x. If the loop crosses the origin, then the polynomial is zero, for some value of x. Q.E.D. (Though you might want to give a moment's thought to the case where the constant term in the polynomial is zero.)

§6.6   The unhappy thing—from an algebraic point of view, I mean—about this proof is that it depends on the matter of *continuity*. I argued that as *x* changes gradually and slowly, so does the corresponding value of the polynomial. This is perfectly true, but it is only true because of the nature of the complex number system, in which you can glide without any jumps or bumps from one number to another, over the dense infinity of numbers in between.

Not all number systems are so accommodating. Number systems are many and various in modern algebra, and we can set up polynomials, and polynomial equations, in all of them. Not many are as friendly as the system of complex numbers, and the FTA is not true in all of them.

From the point of view of modern algebra, therefore, the FTA is a statement about a property of the complex number system, the property known in modern jargon as *algebraic closure*. The system of complex numbers (it says) is algebraically closed—which is to say, any single-unknown polynomial equation with coefficients in the system has a solution in the system. The FTA is not a statement about polynomials, equations, or number systems in general. That is why some mathematicians will take haughty pleasure in telling you that it is not fundamental; and while it is probably a theorem, it is not really a theorem in algebra but in analysis, where the notion of continuity most properly belongs.[60]

Math Primer

ROOTS OF UNITY

§RU.1   IN MY PRIMER ON THE SOLUTION of the general cubic, I mentioned the cube roots of 1 (§CQ.4). There are three of these little devils. Obviously 1 itself is a cube root of 1, since $1 \times 1 \times 1 = 1$. The other two cube roots of 1 are

$$\frac{-1 + i\sqrt{3}}{2} \quad \text{and} \quad \frac{1 - i\sqrt{3}}{2}$$

They are conventionally called ω and ω^2, respectively. If you cube either of these numbers—try it, remembering of course that $i^2 = -1$—you will find that the answer is indeed 1 in either case. Furthermore, the second of these numbers is the square of the first, and the first is the square of the second. ω^2 is of course the square of ω. Only a bit less obviously, ω is the square of ω^2 (because the square of ω^2 is ω^4 which is $\omega^3 \times \omega$ and ω^3 is 1, by definition).

§RU.2   The study of the nth roots of 1—"of unity," we more often say—is very fascinating, and touches on several different areas of math, including classical geometry and number theory. It became possible only when mathematicians were at ease with complex numbers, which is to say from about the middle of the 18th century. The great Swiss mathematician Leonhard Euler broke it wide open in 1751 with a paper titled "On the Extraction of Roots and Irrational Quantities."

The square roots of 1 are of course 1 and −1. The cube roots of 1 are 1 and the two numbers I gave above, ω and ω^2. The fourth roots of 1 are 1, −1, i, and −i. Any one of those will, if you raise it to the fourth power, give you 1. Euler showed that the fifth roots of unity are as follows:

$$1, \quad \frac{\left(-1+\sqrt{5}\right)+i\sqrt{10+2\sqrt{5}}}{4}, \quad \frac{\left(-1-\sqrt{5}\right)+i\sqrt{10-2\sqrt{5}}}{4},$$

$$\frac{\left(-1-\sqrt{5}\right)-i\sqrt{10-2\sqrt{5}}}{4}, \quad \frac{\left(-1+\sqrt{5}\right)-i\sqrt{10+2\sqrt{5}}}{4}$$

Numerically speaking, these work out to: 1, $0.309017 + 0.951057i$, $-0.809017 + 0.587785i$, $-0.809017 - 0.587785i$, $0.309017 - 0.951057i$. If you plot them on the usual complex-number plane, their real parts plotted west-east and their imaginary parts plotted south-north, they look like Figure RU-1.

They are in fact the vertices of a regular pentagon with center at the origin. To put it another way, they lie on the circumference of the unit circle—the circle with radius 1—and they divide that circumference into five equal arcs. If you use Greek words to devise an English term meaning "dividing up a circle," you get "cyclotomic." These complex numbers—points of the complex plane—are called *cyclotomic points*.[61]

§RU.3   Where did all those numbers come from? How do we know that the complex cube roots of 1 are those numbers ω and ω^2 spelled out above? By solving equations, that's how.

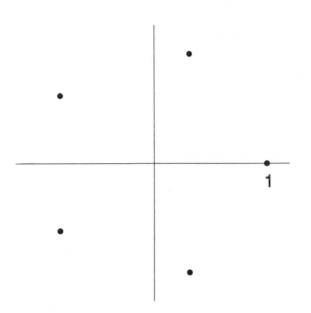

FIGURE RU-1 The fifth roots of unity.

If x is a cube root of 1, then of course $x^3 = 1$. To put it slightly differently, $x^3 - 1 = 0$. But this is just a cubic equation, which we can solve. In fact, since we know that $x = 1$ must be one of the three solutions, we can factorize it right away, to this form:

$$(x - 1)(x^2 + x + 1) = 0$$

So to get the other two roots, we just have to solve that quadratic equation. The solutions are ω and ω^2 just as I described them, from the ordinary quadratic formula (see Endnote 14).

It is generally true, in fact, that the equation whose solutions are the nth roots of unity, the equation $x^n - 1 = 0$, can be factorized like

$$(x - 1) \, (x^{n-1} + x^{n-2} + x^{n-3} + \ldots + x + 1) = 0,$$

and then all the other nth roots of unity—other than 1 itself, I mean—are gotten by solving the equation

$$x^{n-1} + x^{n-2} + \ldots + x + 1 = 0$$

The solution of this equation for general values of n provided 18th- and 19th-century mathematicians with a good deal of employment. Carl Friedrich Gauss gave over a whole chapter of his great 1801 classic *Disquisitiones Arithmeticae* to it—54 pages in the English translation. It is sometimes called the cyclotomic equation of order n, though most modern mathematicians use the term "cyclotomic equation" in a more restricted sense.

§RU.4   The crowning glory of Gauss's investigation was a proof that the regular heptadecagon (that is, a 17-sided polygon) can be constructed in the classical style, using only a ruler and compass.

In the terms in which I have been writing, a regular polygon can be so constructed if and only if the cyclotomic points making up its vertices in the complex plane can be written out with only whole numbers and square root signs. This is the case for $n = 5$, as my display of the fifth roots of unity above show clearly. So a regular pentagon can be constructed with ruler and compass. So, Gauss proved, can a regular heptadecagon. In fact, he wrote out the real part of one of the 17th roots of unity:

$$\frac{-1 + \sqrt{17} + \sqrt{34 - 2\sqrt{17}} + 2\sqrt{17 + 3\sqrt{17} - \sqrt{34 - 2\sqrt{17}} - 2\sqrt{34 + 2\sqrt{17}}}}{16}$$

Nothing but whole numbers and square roots, albeit "nested" three deep, and therefore constructible by ruler and compass. This was the young Gauss's first great mathematical achievement and one so famous that a heptadecagon is inscribed on a memorial at his birthplace of Braunschweig, Germany.

Gauss showed that the same thing is true for any prime (not, as is occasionally said in error, any *number*) having the form $2^{2^k} + 1$. When $k = 0, 1, 2, 3, 4$, this works out to 3, 5, 17, 257, 65537, all prime numbers. When $k = 6$, however, you get 4294967297, which is, "as the distinguished Euler first noticed" (I am quoting Gauss) not a prime number.

§RU.5 The roots of unity have many interesting properties and con
nect not only to classical geometry but to the theory of numbers—of
primes, factors, remainders.

For a glimpse of this, consider the sixth roots of unity. They are 1,
$-\omega^2$, ω, -1, ω^2, and $-\omega$ where ω and ω^2 are the familiar (by now, I
hope) cube roots of unity. If you take each one of these sixth roots in
turn and raise it to its first, second, third, fourth, fifth, and sixth powers, the results are as follows. The roots, denoted by a generic α in the
column headings, are listed down the left-hand column. The first,
second, etc., powers of each root can then be read off along each line.
(The first power of any number is just the number itself. The sixth
power of every sixth root is of course 1.)

α	α^1	α^2	α^3	α^4	α^5	α^6
1:	1	1	1	1	1	1
$-\omega^2$:	$-\omega^2$	ω	-1	ω^2	$-\omega$	1
ω:	ω	ω^2	1	ω	ω^2	1
-1:	-1	1	-1	1	-1	1
ω^2:	ω^2	ω	1	ω^2	ω	1
$-\omega$:	$-\omega^2$	ω^2	-1	ω	$-\omega^2$	1

Only two of the sixth roots, $-\omega^2$ and $-\omega$, generate *all* the sixth
roots in this process. The others just generate some subset of them.
This agrees with intuition, since while $-\omega^2$ and $-\omega$ are *only* sixth roots
of unity, the others are also square roots (in the case of -1) or cube
roots (in the case of ω and ω^2) of unity.

Those nth roots of unity like $-\omega^2$ and $-\omega$ in the case of $n = 6$,
whose successive powers generate *all* the nth roots, are called primitive nth roots of unity.[62] The "first" nth root (proceeding counter-

clockwise around the unit circle in the complex-number diagram) is always primitive. After that the other primitive nth roots are the kth ones, for every number k that has no factor in common with n. The primitive ninth roots of unity, for example, would be the first, second, fourth, fifth, seventh and eighth. If n is a prime number, then every nth root of unity, except for 1, is a primitive nth root of unity. And here we are, as I promised, in number theory, speaking of primes and factors.

Chapter 7

THE ASSAULT ON THE QUINTIC

§7.1   I HAVE DESCRIBED HOW SOLUTIONS to the general cubic and quartic equations were found by Italian mathematicians in the first half of the 16th century. The next obvious challenge was the general quintic equation:

$$x^5 + px^4 + qx^3 + rx^2 + sx + t = 0$$

Let me remind the reader of what is being sought here. For any particular quintic equation, a merely *numerical* solution can be found to whatever degree of approximation is desired, using techniques familiar to the Muslim mathematicians of the 10th and 11th centuries. What was not known was an *algebraic* solution, a solution of the form

$$x = [\text{some algebraic expression in } p, q, r, s, \text{and } t],$$

where the word "algebraic" in those brackets means "involving only addition, subtraction, multiplication, division, and the extraction of roots (square root, cube root, fourth root, fifth root, . . .)." I had better specify, for the sake of completeness, that the bracketed expression should contain only a *finite* number of these operations. The expression for a solution of the general quartic, given in §CQ.7, is the kind of thing being sought here.

We now know that no such solution exists. So far as I can discover, the first person who believed this to be the case—who believed, that is, that the general quintic has no algebraic solution—was another Italian, Paolo Ruffini, who arrived at his belief near the very end of the 18th century, probably in 1798. He published a proof the following year. (Gauss recorded the same opinion in his doctoral dissertation of that same year, 1799, but offered no proof.) Then Ruffini published second, third, and fourth proofs in 1803, 1808, and 1813. None of these proofs satisfied his fellow mathematicians, insofar— and it was not very far—that they paid any attention to them. Credit for conclusively proving the nonexistence of an algebraic solution to the general quintic is generally given to the Norwegian mathematician Niels Henrik Abel, who published his proof in 1824.

For practically the entire 18th century, therefore, it was believed that the general quintic equation had an algebraic solution. To find that solution, though, must have been reckoned a problem of the utmost difficulty. By 1700 it had, after all, been 160 years since Lodovico Ferrari had cracked the quartic, and no progress on the quintic had been made at all. The kinds of techniques used for the cubic and the quartic had not scratched the quintic. Plainly some radically new ideas were needed.

The matter fell into neglect during the 17th and early 18th centuries. With the powerful new literal symbolism now available, the discovery of calculus, the domestication of complex numbers, and the accelerating growth of the theoretical sciences, there was a great deal of low-hanging fruit for mathematicians to pick. Well-tried and apparently intractable problems of no obvious practical application tend to lose their appeal under such circumstances. This was, remember, a quintessentially—if you will pardon the expression— pure-mathematical problem. Anyone who needed an actual numerical solution to an actual quintic equation could easily find one.

§7.2 The great Swiss mathematician Leonhard Euler (pronounced "oiler") first tackled the problem of the general quintic in 1732 while living in St. Petersburg, Russia. He did not get very far with it on that occasion, but 30 years later, while working for Frederick the Great in Berlin, he had another go. In this paper ("On the Solution of Equations of Arbitrary Degree"), Euler suggested that an expression for the solutions of an nth-degree equation might have the form

$$A + B\sqrt[n]{\alpha} + C\left(\sqrt[n]{\alpha}\right)^2 + D\left(\sqrt[n]{\alpha}\right)^3 + \ldots$$

where α is the solution of some "helper" equation of degree $n-1$, and $A, B, C, \ldots$ are some algebraic expressions in the original equation's coefficients. Well and good, but how do we know that this helper equation of degree $n-1$ can be found?

We don't, and with that the greatest mathematical mind of the 18th century[63] (counting Gauss as belonging to the 19th) let matters stand. Euler's work was not without result, though. Abel's 1824 proof of the unsolvability of the general quintic opens with a form for the solutions very much like that last expression above.

§7.3 In the odd way these things sometimes happen, the crucial insights first came not from the 18th century's greatest mathematician but from one of its least.

Alexandre-Théophile Vandermonde, a native Frenchman in spite of his name, read a paper to the French Academy[64] in Paris in November 1770, when he was 35 years old. He subsequently read three more papers to the academy (to which he was elected in 1771), and these four papers were his entire mathematical output. His main life interest seems to have been music. Says the *DSB*: "It was said at that time that musicians considered Vandermonde to be a mathematician and that mathematicians viewed him as a musician."

Vandermonde is best known for a determinant named after him (I shall discuss determinants later); yet the determinant in question does not actually appear in his work, and the attribution of it to him seems to be a misunderstanding. Vandermonde is altogether an odd, shadowy figure, like one of Vladimir Nabokov's inventions. Later in life he became a Jacobin, an ardent supporter of the French Revolution, before his health failed and he died in 1796.

Vandermonde's key insight was a way of writing *each* solution of an equation in terms of *all* the solutions. Consider, for example, the quadratic equation $x^2 + px + q = 0$. Suppose its solutions are α and β. The following two things are obviously true:

$$\alpha = \frac{1}{2}\left[(\alpha + \beta) + (\alpha - \beta)\right]$$

$$\beta = \frac{1}{2}\left[(\alpha + \beta) - (\alpha - \beta)\right]$$

To put it slightly differently: If you consider the square root sign to indicate two possible values, one plus and one minus, then the solutions of the quadratic equation are given by the two possible values of this expression:

$$\frac{1}{2}\left[(\alpha + \beta) + \sqrt{(\alpha - \beta)^2}\right]$$

What is the point of that? Well, $(\alpha - \beta)^2$ is equal to $\alpha^2 - 2\alpha\beta + \beta^2$, which is equal to $(\alpha + \beta)^2 - 4\alpha\beta$, and that is a *symmetric polynomial* in α and β. Symmetric polynomials in the solutions, remember, can always be translated into polynomials in the coefficients, in this particular case $p^2 - 4q$.

From this comes the familiar solution to the general quadratic equation. That, of course, is not the main point. The main point is that this approach looks like one that can be generalized to an equation of any degree, whereas previous approaches to the solutions of the quadratic, cubic, and quartic were all ad hoc and not generalizable.

Let's generalize the above procedure to the depressed cubic equation $x^3 + px + q = 0$. Using ω and ω^2 for the complex cube roots of unity as usual and recalling from §RU.3 that they satisfy the quadratic equation $1 + \omega + \omega^2 = 0$, I can write my general solution as

$$\frac{1}{3}\left[(\alpha+\beta+\gamma)+\sqrt[3]{(\alpha+\omega\beta+\omega^2\gamma)^3}+\sqrt[3]{(\alpha+\omega^2\beta+\omega\gamma)^3}\right]$$

Since this is a depressed cubic, $(\alpha + \beta + \gamma)$ is zero (see §5.6), and we need only bother with the two cube root terms. It seems that there is a drawback here, though: The expressions under the cube root signs are not symmetric in $\alpha, \beta,$ and γ as the one under the square root sign was symmetric in α and β when I tackled the quadratic equation.

Let me investigate that a little more closely. I'll put $U = (\alpha + \omega\beta + \omega^2\gamma)^3$, and $V = (\alpha + \omega^2\beta + \omega\gamma)^3$. What happens to U and V under the six possible permutations of $\alpha, \beta,$ and γ? Well, denoting the permutations in a way that I hope is obvious:

Permutation	U	V
$\alpha \to \alpha, \beta \to \beta, \gamma \to \gamma$:	$(\alpha + \omega\beta + \omega^2\gamma)^3$	$(\alpha + \omega^2\beta + \omega\gamma)^3$
$\alpha \to \alpha, \beta \to \gamma, \gamma \to \beta$:	$(\alpha + \omega^2\beta + \omega\gamma)^3$	$(\alpha + \omega\beta + \omega^2\gamma)^3$
$\alpha \to \gamma, \beta \to \beta, \gamma \to \alpha$:	$(\omega^2\alpha + \omega\beta + \gamma)^3$	$(\omega\alpha + \omega^2\beta + \gamma)^3$
$\alpha \to \beta, \beta \to \alpha, \gamma \to \gamma$:	$(\omega\alpha + \beta + \omega^2\gamma)^3$	$(\omega^2\alpha + \beta + \omega\gamma)^3$
$\alpha \to \beta, \beta \to \gamma, \gamma \to \alpha$:	$(\omega^2\alpha + \beta + \omega\gamma)^3$	$(\omega\alpha + \beta + \omega^2\gamma)^3$
$\alpha \to \gamma, \beta \to \alpha, \gamma \to \beta$:	$(\omega\alpha + \omega^2\beta + \gamma)^3$	$(\omega^2\alpha + \omega\beta + \gamma)^3$

(That first nothing-happens permutation is the "identity permutation.")

This doesn't look very informative. Remember, though, that ω is a cube root of unity. We can use this fact to "pull out" the ω and ω^2 from in front of the α terms. For example, taking the first term in the fifth row:

$$\left(\omega^2\alpha+\beta+\omega\gamma\right)^3 \;=\; \left[\omega^2\left(\alpha+\omega\beta+\omega^2\gamma\right)\right]^3 \;=\; \omega^6\left(\alpha+\omega\beta+\omega^2\gamma\right)^3$$

which is just U. In fact, every one of those permutations works out to deliver either U or V. Half work out to U, half to V. The effect on U and V of any possible permutation of the solutions is either to leave U as U and V as V or to exchange U with V.

Let me just state that again, slightly differently, for effect. I tried all possible permutations of the solutions and found that half the permutations left U and V unchanged; the other half turned U into V and V into U.

The key concept here is *symmetry*. A polynomial in the solutions α, β, and γ may be totally symmetric, like the ones Viète and Newton investigated: Permute the solutions in all six possible ways and the value of the polynomial won't change. It only has one value. Or a polynomial might be totally asymmetric: Permute the solutions in all six possible ways and the value of the polynomial will take six different values. Or, as in my example, the polynomial may be *partially* symmetric: Permute the solutions in all six possible ways and the value of the polynomial will take some number of values greater than 1 but less than 6.

Proceeding now to the solution: It follows from all this that any possible permutation of the solutions α, β, and γ will leave $U + V$ and UV (or any other symmetric polynomial in U and V) unchanged. So $U + V$ and UV must themselves be symmetric polynomials in the solutions α, β, and γ, and therefore they can be expressed in terms of the coefficients p and q of the cubic. In fact, if you chew through the algebra (and remember again that for a depressed cubic, $\alpha + \beta + \gamma$, the coefficient of the x^2 term, is zero), you will get

$$U + V = -27q, \quad UV = -27p^3$$

So if you solve the quadratic equation

$$t^2 + 27qt - 27p^3 = 0$$

for the unknown t, you shall have U and V. You have then solved the cubic. (Compare this approach with the one in §CQ.6.)

There is a second drawback to this method. As before, the root sign in that expression for a general solution is understood to embrace all possible values of the root—in this case, all three possible values of the cube root: a number, ω times that number, and ω^2 times that number. Since there are two cube roots in my square brackets, the expression represents nine numbers altogether: the three solutions of my cubic and six other irrelevant numbers. How do I know which are which?

Vandermonde did not really overcome this problem. He had, though, introduced the key insight. In terms of the cubic:

> Write a general solution in terms of a symmetric, or partially symmetric, polynomial in *all* the solutions.
>
> Ask: How many different values can this polynomial take under all six possible permutations of $\alpha, \beta,$ and γ?
>
> The answer is two, the ones I called U and V; this fact leads us to a quadratic equation.

This was the first attempt at solving equations by looking at the permutations of their solutions and at a *subset* of those permutations that left some expression—the cube of $\alpha + \omega\beta + \omega^2\gamma$ in my example—unchanged. These were key ideas in the attack on the general quintic.

§7.4   Alas for Vandermonde, his work was completely overshadowed by a much greater talent. Says Professor Edwards: "Unlike Vandermonde, who was French but did not have a French name, Lagrange had a French name but was not French."[65]

Giuseppe Lodovico Lagrangia was born in Turin, in northwest Italy just 30 miles from the French border, in 1736. Though not French, he was of part-French ancestry and seems to have preferred writing in French, using the French form of his surname from an early age. (He spoke French with a strong Italian accent all his life, however.) He became a member of the French Academy in 1787 and spent the rest of his life in Paris, dying there in 1813. He weathered the French Revolution and was instrumental in setting up the metric system of weights and measures. It is therefore not unjust that he is known to us as Joseph-Louis Lagrange. There is a pretty little park in Paris named after him; it contains the city's oldest tree.[66]

Lagrange's early working life was spent in Turin, where he was a professor of mathematics at age 16. By the time Vandermonde was presenting his paper to the French Academy in 1770, Lagrange had moved to Berlin, to the court of Frederick the Great. He was in fact Euler's successor at that court, having arrived in 1766 as Euler left. Frederick apparently found Lagrange, who was well read in contemporary politics and philosophy and had a sly, ironic style of wit, much more *gemütlich* than the no-frills Euler and pronounced himself delighted with the change.

Lagrange's great contribution to algebra came in 1771, a few months after Vandermonde's presentation to the Academy in Paris. It was published by Frederick the Great's own Academy, in Berlin, as a paper titled "Reflections on the Algebraic Solution of Equations." It was this paper, by an already distinguished mathematician, that brought ideas about approaching equations via permutations of their solutions to the front of mathematicians' minds.

It is all very unfair. Vandermonde thought of this first and is duly acknowledged in modern textbooks. His paper, however, went unnoticed and, according to Bashmakova and Smirnova, "had no effect on the evolution of algebra." It was not even published until 1774, by which time Lagrange's paper had been widely circulated. There is no evidence that Lagrange knew of Vandermonde's work. He was not, in any case, a devious man and would have acknowledged that work if he had known about it. It was just a case of great minds—or more accurately, a great mathematical mind and a good one—thinking alike.

Lagrange followed the same train of thought as Vandermonde, but he was a stronger mathematician and took the argument further.

I shall stick with the general depressed cubic equation $x^3 + px + q = 0$ by way of illustration.

Beginning with the same expression that Vandermonde had used, $\alpha + \omega\beta + \omega^2\gamma$ (it is technically known as the *Lagrange resolvent*), Lagrange noted that this takes six different values when you permute the solutions α, β, and γ, though its cube takes only two, as I showed above. The values are

$$t_1 = \alpha + \omega\beta + \omega^2\gamma \qquad t_2 = \alpha + \omega^2\beta + \omega\gamma \qquad t_3 = \omega^2\alpha + \omega\beta + \gamma$$
$$t_4 = \omega\alpha + \beta + \omega^2\gamma \qquad t_5 = \omega^2\alpha + \beta + \omega\gamma \qquad t_6 = \omega\alpha + \omega^2\beta + \gamma$$

As before, I can "pull out" the omegas from the α terms, using the fact that ω is a cube root of 1. Then, $t_3 = \omega^2 t_2$, $t_4 = \omega t_2$, $t_5 = \omega^2 t_1$, and $t_6 = \omega t_1$.

Now form the sixth-degree polynomial that has the t's as its solutions. This will be

$$(X - t_1)(X - t_2)(X - \omega^2 t_2)(X - \omega t_2)(X - \omega^2 t_1)(X - \omega t_1)$$

Lagrange calls it the *resolvent equation*. It easily simplifies to

$$(X^3 - t_1^{\;3})(X^3 - t_2^{\;3}),$$

which is the quadratic equation we got before, with solutions $U = t_1^{\;3}$ and $V = t_2^{\;3}$.

Lagrange carried out the same procedure for the general quartic, this time getting a resolvent equation of degree 24. Just as the resolvent for the cubic, though of degree 6, "collapsed" into a quadratic, so the degree-24 resolvent for the quartic collapses into a degree-6 equation. That looks bad, but the degree-6 equation in X turns out to actually be a cubic equation in X^2, and so can be solved.

The number of possible permutations of five objects is $1 \times 2 \times 3 \times 4 \times 5$, which is 120. Lagrange's resolvent equation for the quintic therefore has degree 120. With some ingenuity, this can be collapsed into an equation of degree 24. There, however, Lagrange got stuck. He had, though, like Vandermonde, grasped the essential point: In order to understand the solvability of equations, you have to investigate the permutations of their solutions and what happens to certain key expressions—the resolvents—under the action of those permutations.

He had also proved an important theorem, still taught today to students of algebra as Lagrange's theorem. I shall state it in the terms in which Lagrange himself understood it. The modern formulation is quite different and more general.

Suppose you have a polynomial[67] in n unknowns. There are $1 \times 2 \times 3 \times \ldots \times n$ ways to permute these unknowns. This figure, as you probably know, is called "the factorial of n" and is written with an exclamation point: $n!$ So, $2! = 2$, $3! = 6$, $4! = 24$, $5! = 120$, and so on. (The value of $1!$ is conventionally taken to be 1. So, for deep but strong reasons, is $0!$) Suppose you switch around the unknowns in all $n!$ possible ways, as I did with α, β, and γ in the previous section. How many different values will the polynomial have? The answer in that previous section was 2, the values I called U and V. But what, if anything, can be said about the answer *in general?* If some polynomial takes A different values, is there anything we can say for sure about A?

Lagrange's theorem says that A will always be some number that divides $n!$ exactly. So form any polynomial you like in α, β, and γ, shuffle the three unknowns in all six possible configurations, and tally how many different values your polynomial takes. The answer will be 1, if your polynomial is symmetric. It may be 2, as was the case in my example up above. It may be 3, as in the case of this polynomial: $\alpha + \beta - \gamma$. It may be 6, as with this one: $\alpha + 2\beta + 3\gamma$. However, it will never be 4 or 5. That's what Lagrange proved—though, of course, he proved it for any number n, not just $n = 3$.

(Note that Lagrange's theorem tells you a property of A: that it will divide $n!$ exactly. It does not guarantee that *every* number that divides $n!$ exactly is a possible A—a possible number of values for some polynomial to take on under the $n!$ permutations. Suppose n is 5, for instance. Then $n!$ is 120. Since 4 divides exactly into 120, you might expect that there exists some polynomial in five unknowns which, if you run the unknowns through all 120 possible permutations, takes on four values. This is not so. The fact of its not being so was discovered by Cauchy, of whom I shall say more in the next section, and is critical to the problem of finding an algebraic solution to the general quintic.)

Lagrange's theorem is one of the cornerstones of modern group theory—a theory that did not even exist in Lagrange's time.

§7.5   The first name in this chapter was that of Paolo Ruffini, author of several attempts to prove there is no algebraic solution of the general quintic. Ruffini followed Lagrange's ideas. For the general cubic equation we can get a resolvent equation that is quadratic, which we know how to solve. For a quartic equation, we can get a resolvent equation that is cubic, and we know how to solve cubics, too. Lagrange had shown that for an algebraic solution of the general quintic equation, we need to devise a resolvent equation that is cubic or quartic. Ruffini, by a close scrutiny of the values a polynomial can take when you permute its unknowns, showed that this was impossible.

"One has to feel desperately sorry for Ruffini," remarks one of his biographers.[68] Indeed one does. His first proof was flawed, but he kept working on it and published at least three more. He sent these proofs off to senior mathematicians of his day, including Lagrange, but was either ignored or brushed off with uncomprehending condescension—which must have tasted especially bitter coming from Lagrange, whom Italians considered a compatriot. Ruffini submitted his proofs to learned societies, including the French Institute (a replacement for the Academy, which had been temporarily abolished by the Revolution) and Britain's Royal Society. The results were the same.

Almost until he died in 1822, poor Ruffini tried without success to get recognition for his work. Only in 1821 did any real acknowledgment come to him. In that year the great French mathematician Augustin-Louis Cauchy sent him a letter, which Ruffini must have treasured in the few months he had left. The letter praised his work and declared that in Cauchy's opinion Ruffini had proved that the general quintic equation has no algebraic solution. In fact, Cauchy had produced a paper in 1815 clearly based on Ruffini's.[69]

I should pause here to say a word or two about Cauchy, since his name will crop up again in this story. It is a great name in the history of mathematics. "More concepts and theorems have been named for Cauchy than for any other mathematician (in elasticity alone there are sixteen concepts and theorems named for Cauchy)." That is from Hans Freudenthal's entry on Cauchy in the *DSB*—an entry that covers 17 pages, the same number of pages as the entry for Gauss.

Cauchy's style of work was very different from Gauss's, though. Gauss published sparingly, only making known those results he had worked and polished to perfection. (This is why his publications are nearly unreadable.) It has been a standing joke with mathematicians for 150 years that when one has come up with a brilliant and apparently original result, the first task is to check that it doesn't appear in Gauss's unpublished papers somewhere. Cauchy, by contrast, published everything that came into his head, often within days. He actually founded a private journal for this purpose.

Cauchy's personality has also generated much comment. Different biographers have drawn him as a model of piety, integrity, and charity, or as a cold-blooded schemer for power and prestige, or as an idiot savant, blundering through life in a state of unworldly confusion. He was a devout Catholic and a staunch royalist—a reactionary in a time when Europe's intellectual classes were beginning to take up their long infatuation with secular, progressive politics.[70]

Modern commentary has tended to give Cauchy the benefit of the doubt on many issues once thought damning (though see §8.6). Even E. T. Bell deals evenhandedly with Cauchy: "His habits were temperate and in all things except mathematics and religion he was moderate." Freudenthal inclines to the idiot-savant opinion: "[H]is quixotic behavior is so unbelievable that one is readily inclined to judge him as being badly melodramatic. . . . Cauchy was a child who was as naïve as he looked." Whatever the facts of the man's personality, that he was a very great mathematician cannot be doubted.

§7.6   Under the circumstances, then, we should pity poor Ruffini and look with scorn on Niels Henrik Abel, to whom credit is commonly given for proving the algebraic unsolvability of the general quintic.

In fact, nobody thinks like that. For one thing, Cauchy's opinion notwithstanding, mathematicians of his own time thought Ruffini's proofs were flawed. (Modern views have been kinder to Ruffini, and the algebraic unsolvability of the general quintic is now sometimes called the Abel-Ruffini theorem.) For another the proofs were written in a style difficult to penetrate—this was Lagrange's main problem with them. And for another, Abel is a person whose life presents a much more pitiable spectacle than Ruffini's, though he seems to have been a cheerful and sociable man despite it all.

Abel was the first of the great trio of 19th-century Norwegian algebraists. We shall meet the other two later. He came from a place on the northern windswept fringes of Europe—near Stavanger, on the "nose" of Norway—poor in itself and made poorer and unhappier by the instability of the times. His own family belonged to the genteel poor. His father and grandfather were both country pastors. Abel's father fell into political misfortune, took to drink, and "died an alcoholic, leaving nine children and a widow who also turned to alcohol for solace. After his funeral, she received visiting clergy while in bed with her peasant paramour."[71]

For the rest of his short life—he died a few weeks before his 27th birthday—Abel was chronically hard up at the best of times and deep in debt at the worst. His country's condition was similar. By the time Abel reached his teens, Norway was semi-independent as part of the joint kingdoms of Norway and Sweden, with a capital at Oslo, then called Christiania,[72] and a parliament of her own, but living in the economic and military shadow of the richer, more populous Sweden. It is to the Norwegians' great credit that they scraped together enough funds to send this unknown young mathematician on a European tour from 1825 to 1827, though the meagerness of those funds and the vigilance with which their expenditure was supervised has aroused indignation in some of Abel's biographers and seems to have inspired guilt in later Norwegian governments.

Abel had discovered mathematics very early and had had the good fortune to come under the guidance of a teacher, Bernt Michael Holmboë, who recognized his talent and who, though not a creative mathematician himself, knew his way around the major texts of the time. With Holmboë's encouragement and financial help, Abel attended the new University of Christiania in 1821–1822.

Abel had already been working on the problem of the general quintic since 1820, had proved the unsolvability theorem, and in 1824 had paid from his own pocket to have the proof printed up. To save on expenses he condensed it to just six pages, sacrificing much of the proof's coherence in the process. Still, he felt sure that those six pages would open the doors of Europe's greatest mathematicians to him.

That was, of course, not quite what happened. The great Gauss, presented with a copy of Abel's proof in advance of a visit from Abel in person, tossed it aside in disgust. This is not quite as shameful as it sounds. Gauss was already famous, and famous mathematicians, then as now, suffer considerably from the attention of cranks with claims to have proved some outstanding problem or other.[73] Gauss was not a person who suffered fools gladly at the best of times, and he seems to have had little interest in the algebraic solution of polynomial equations. Abel canceled the planned visit to Gauss.

To make up for this disappointment, Abel had a great stroke of luck in Berlin. He met August Crelle, a unique figure in the history of math. Crelle—his dates are 1780–1855—was not a mathematician, but he was a sort of impresario of math. He had a keen eye for mathematical talent and excellence and, when he found it, did what he could to nourish it. Crelle was a self-made man and largely self-educated, too, from humble origins. He got a job as a civil engineer with the government of Prussia and rose to the top of that profession. He was in part responsible for the first railroad in Germany, from Berlin to Potsdam, 1838. Sociable, generous, and energetic, Crelle played the barren midwife to great mathematical talents. He made a huge contribution to 19th-century math, though indirectly.

At just the time when Abel arrived in Berlin in 1825, Crelle had made up his mind to found his own mathematics journal. Crelle spotted the young Norwegian's talent (they apparently conversed in French), introduced him to everyone in Berlin, and published his unsolvability proof in the first volume of his *Journal of Pure and Applied Mathematics* in 1826. He published many more of Abel's papers, too. The unsolvability of the quintic was merely one aspect, a minor one, of Abel's wide-ranging mathematical interests. His major work was in analysis, in the theory of functions.

Abel returned penniless to Christiania in the spring of 1827 and died in 1829 from tuberculosis, the great curse of that age, without having left Norway again. Two days later, of course not knowing of Abel's death, Crelle wrote to tell him that the University of Berlin had offered him a professorship.

Abel's proof is assembled from ideas he picked up from Euler, Lagrange, Ruffini, and Cauchy, all mortared together with great ingenuity and insight. Its general form is of the type called *reductio ad absurdum*: that is, he begins by assuming the opposite of what he wants to prove and shows that this implies a logical absurdity.

What Abel wants to show is that there is no algebraic solution to the general quintic. He therefore begins by assuming that there *is* such a solution. He writes his general quintic like this:

$$y^5 - ay^4 + by^3 - cy^2 + dy - e = 0$$

Then he says: OK, let's say there is an algebraic solution, all the solutions y being represented by expressions in a, b, c, d, and e, these expressions involving only a finite number of additions, subtractions, multiplications, divisions, and extractions of roots. Of course, the root extractions might be "nested," like the square roots under the cube roots in the solution of the general cubic. So let's express the solutions in some general yet useful (for our purposes) way that allows for this nesting, in terms of quantities that might have roots under them, and under *them*, and under *them* . . . .

Abel comes up with an expression for a general solution that closely resembles the one of Euler's that I showed in §7.2. Borrowing from Lagrange (and Vandermonde, though Abel did not know that), he argues that this general solution must be expressible as a polynomial in *all* the solutions, along with fifth roots of unity, as in my §§7.3–7.4. Abel then picks up on the result of Cauchy's that I mentioned at the end of §7.4: A polynomial in five unknowns can take two different values if you permute the unknowns, or five different values, but not three or four. Applying this result to his general expression for a solution, Abel gets his contradiction.[74]

§7.7   Abel's proof—or, if you want to be punctilious about it, the Abel-Ruffini proof—that the general quintic has no algebraic solution, closes the first great epoch in the history of algebra.

In writing of the closing of an epoch, I am imposing hindsight on the matter. Nobody felt that way in 1826. In fact, Abel's proof took some time to become widely known. Nine years after its publication, at the 1835 meeting of the British Association in Dublin, the mathematician G. B. Jerrard presented a paper in which he claimed to have found an algebraic solution to the general quintic! Jerrard was still pressing his claim 20 years later.

Nor did Abel's proof put an end to the general theory of polynomial equations in a single unknown. Even though there is no algebraic solution to the *general* quintic, we know that *particular* quintics have solutions in roots. In my primer on roots of unity, for example, I showed exactly such a solution for the equation $x^5 - 1 = 0$, which is indisputably a quintic. So the question then arises: *Which* quintic equations can be solved algebraically, using just $+, -, \times, \div$, root signs, and polynomial expressions in the coefficients? A complete answer to this question was given by another French mathematical genius, Évariste Galois, whose story I shall tell in Chapter 11.

There was, though, a great, slow shift in algebraic sensibility in the early decades of the 19th century. It had been under way for some time, since well before Abel printed up his six-page proof. I have characterized this new way of thinking as "the discovery of new mathematical objects." Through the 18th century and into the beginning of the 19th, algebra was taken to be what the title of Newton's book had called it: universal arithmetic. It was arithmetic—the manipulation of numbers—by means of symbols.

All through those years, however, European mathematicians had been internalizing the wonderful new symbolism given to them by the 17th-century masters. Gradually the attachment of the symbols to the world of numbers loosened, and they began to drift free, taking on lives of their own. Just as two numbers can be added to give a new number, were there not other kinds of things that might be combined, two such being merged together to give another instance of the same kind? Certainly there were. Gauss, in his 1801 classic *Disquisitiones Arithmeticae*, had dealt with *quadratic forms*, polynomials in two unknowns, like this:

$$AX^2 + 2BXY + CY^2$$

His investigations led him to the idea of the *composition* of such forms, a way of melding two forms together to get a new one, more subtle than simple addition or multiplication of expressions. This was, wrote Gauss, "a topic no one has yet considered."

And then there was Cauchy's 1815 memoir on the number of values a polynomial can take when its unknowns are permuted. This was the memoir Abel had used in his proof. In it Cauchy introduces the idea of *compounding permutations*.

To give a simple illustration: Suppose I have three unknowns α, β, and γ, and suppose I refer to the switching of β and γ as permutation X, the switching of α and β as permutation Y. Now suppose I first do permutation X and then permutation Y. What has happened? Well, permutation X turned (α, β, γ) into (α, γ, β). Then permutation Y turned this into (β, γ, α). So the net effect is to turn (α, β, γ) into (β, γ, α), which is another permutation! We could call it permutation Z, and speak of compounding permutation X and permutation Y to get permutation Z. That is in fact exactly how Cauchy did describe his manipulations, thereby essentially inventing the theory of groups (though he did not use that term).

In doing so Cauchy entered a strange new world. Note, for example, how the analogy between compounding permutations and adding numbers breaks down in an important respect. If you first do permutation Y and then permutation X, the result is (γ, α, β). So it matters which order you do your compounding in. This is not the case with addition of numbers, in which 7 + 5 is equal to 5 + 7. This particular property of ordinary numbers is technically known as *commutativity*. Cauchy's compounding of permutations was non-commutative.

All of this was in the air in the early 19th century. After half a dozen generations of working with the literal symbolism of Viète and Descartes, mathematicians were beginning to understand that the compounding of numbers by addition and multiplication to get other numbers is only a particular case of a kind of manipulation that can be applied much more widely, to objects that need not be numbers at all. Those symbols they had gotten so used to might stand for *anything*: numbers, permutations, arrays of numbers, sets, rotations, transformations, propositions, . . . When this sank in, modern algebra was born.

§7.8 For the next few chapters I shall set aside a strictly chronological approach to my narrative. I have come to a period—the middle two quarters of the 19th century—of great fecundity in new algebraic ideas. Not only were groups discovered in those years, but also many other new mathematical objects. "Algebra" ceased to be only a singular noun, and became a plural. The modern concepts of "field," "ring," "vector space," and "matrix" took form. George Boole brought logic under the scope of algebraic symbolism, and geometers found that, thanks to algebra, they had many more than three dimensions to explore.

The historian of these fast-moving developments has two choices. He can stick to a strictly chronological scheme, trying to show how new ideas came up and interacted with others year by year, or he can follow one single train of thought through the period, then loop back and pick up another. I am going to take the latter approach, making several passes at this period of tumultuous growth in algebra and of radical changes in algebraic thinking. First, a trip into the fourth dimension.

Math Primer

VECTOR SPACES AND ALGEBRAS

§VS.1 THE HISTORY OF THE CONCEPT *vector* in mathematics is rather tangled. I shall try to untangle it in the main text. What follows in this primer is an entirely modern treatment, developed in hindsight, using ideas and terms that began to be current around 1920.

§VS.2 *Vector space* is the name of a mathematical object. This object has two kinds of elements in it: *vectors* and *scalars*. The scalars are probably some familiar system of numbers, with full addition, subtraction, multiplication, and division—$\mathbb{R}$ will do just fine. Vectors are a little more subtle.

Let me give a very simple example of a vector space. Consider an infinite flat plane. I select one particular point of the plane, which I call the *origin*. A vector is a line going from the origin to some other point. Figure VS-1 shows some vectors. You can see that the two characteristics of a vector are its *length* and its *direction*.

Every vector in this vector space has an inverse. The inverse is a line of the same length as the vector but pointing *in the opposite direction* (see Figure VS-2). The origin by itself is counted as a vector, called the *zero vector*.

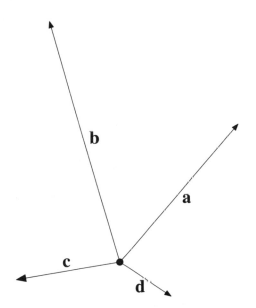

FIGURE VS-1 Some vectors.

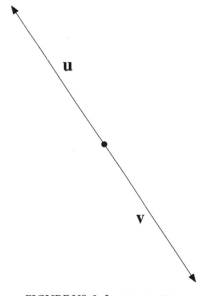

FIGURE VS-2 Inverse vectors.

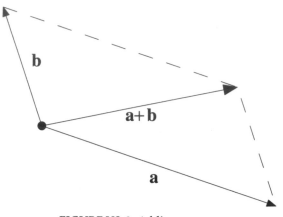

FIGURE VS-3 Adding vectors.

Any two vectors can be added together. To add two vectors, look at them as adjacent sides of a parallelogram. Sketch in the other two sides of this parallelogram. The diagonal proceeding from the origin to the far corner of the parallelogram is the sum of the two vectors. (see Figure VS-3).

If you add a vector to the zero vector, the result is just the original vector. If you add a vector to its inverse, the result is the zero vector.

Any vector can be multiplied by any scalar. The length of the vector changes by the appropriate amount (for example, if the scalar is 2, the length of the vector is doubled), but its direction is unchanged—except that it is precisely reversed when the multiplying scalar is negative.

§VS.3   That's pretty much it. Of course, the flat-plane vector space I have given here is just an illustration. There is more to vector spaces than that, as I shall try to show in a moment. My illustration will serve for a little longer yet, though.

An important idea in vector space theory is *linear dependence.*
Take any collection of vectors in your vector space, say **u**, **v**, **w**, . . . etc.
If it is possible to find some scalars p, q, r, . . . , not all zero, such that

$$p\mathbf{u} + q\mathbf{v} + r\mathbf{w} + \ldots = \mathbf{0} \text{ (that is, the zero vector)},$$

then we say that **u**, **v**, **w**, . . . are *linearly dependent.* Look at the two
vectors **u** and **v** in Figure VS-4. **v** points in precisely the opposite
direction to **u**, and is $\frac{2}{3}$ its length. It follows that $2\mathbf{u} + 3\mathbf{v} = \mathbf{0}$. So **u** and
v are linearly dependent. Another way of saying this is: You can ex-
press one of them in terms of the other: $\mathbf{v} = -\frac{2}{3}\mathbf{u}$.

Now look at the vectors **u** and **v** in Figure VS-5. This time the two
vectors are not linearly dependent. You cannot possibly find any sca-
lars a and b, not both zero, making $a\mathbf{u} + b\mathbf{v}$ equal to the zero vector.
(A good way to convince yourself of this is to argue that since a and b
can't both be zero, we can divide through that expression by one of

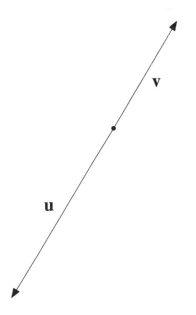

FIGURE VS-4 Linear dependence.

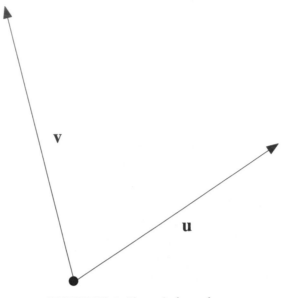

FIGURE VS-5 Linear independence.

them—say b—to get it in this form: $c\mathbf{u} + \mathbf{v}$. Now imagine the diagram for adding $\mathbf{u}$ and $\mathbf{v}$, with the parallelogram and its diagonal drawn in—Figure VS-3 above. Leaving $\mathbf{v}$ alone, change the size of the $\mathbf{u}$ vector by varying c through all possible values, making it as big as you like, as small as you like, zero, or negative—reversing its direction for negative values of c, of course—and watch what happens to the diagonal. *It can never be zero.*)

The vectors $\mathbf{u}$ and $\mathbf{v}$ in Figure VS-5 are therefore not linearly dependent. They are linearly *independent.* You can't express one in terms of the other. In Figure VS-4, you *can* express one in terms of the other: $\mathbf{v}$ is $-\frac{2}{3}\mathbf{u}$.

Equipped with the idea of a linearly independent set of vectors, we can define the *dimension* of a vector space. It is the largest number of linearly independent vectors you can find in the space. In my sample vector space, you can find plenty of examples of two linearly independent vectors, like the two in Figure VS-5; but you can't find three. The vector **w** in Figure VS-6 can be expressed in terms of **u** and **v**, in fact the way I have drawn it: **w** = 2**u** – **v**. To put it another way, 2**u** – **v** – **w** = **0**. The three vectors **u**, **v**, and **w** are linearly dependent.

Since the largest number of linearly independent vectors I can find in my sample space is two, this space is of dimension two. I guess this is not a great surprise.

In a two-dimensional vector space like this one, if two vectors are linearly independent, they will not lie in the same line. In a space of three dimensions, if three vectors are linearly independent, they will not lie in the same flat plane. Contrariwise, *three* vectors all lying in the same flat plane—the same flat *two*-dimensional space—will be linearly dependent, just as *two* vectors lying in the same line—the same *one*-dimensional space—will be linearly dependent. Any *four* vectors all in the same *three*-dimensional space will be linearly dependent . . . and so on.

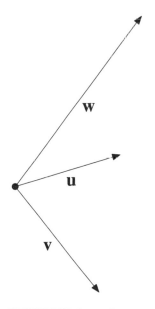

FIGURE VS-6 **w** = 2**u** – **v**.

Given two linearly independent vectors, like **u** and **v** in Figures VS-5 and VS-6, any other vector in my sample space can be expressed in terms of them, as I did with **w**. This means that they are a *basis* for the space. Any two linearly independent vectors will do as a basis; any other vector in the space can be expressed in terms of them.

Dimension and *basis* are fundamental terms in vector space theory.

§VS.4   The concept of a vector space is purely algebraic and need not have any geometric representation at all.

Consider all polynomials in some unknown quantity x, of degree not more than 5, with coefficients in $\mathbb{R}$. Here are some examples:

$$2x^5 - 8x^4 - x^3 + 11x^2 - 9x + 15$$
$$44x^5 + 19x^3 + 4x + 1$$
$$x^2 - 2x + 1$$

These polynomials are my vectors. My scalars are just $\mathbb{R}$. This is a vector space. Look, I can add any two vectors to get another vector:

$$(2x^5 - 8x^4 - x^3 + 11x^2 - 9x + 15) + (44x^5 + 19x^3 + 4x + 1)$$
$$= 46x^5 - 8x^4 + 18x^3 + 11x^2 - 5x + 16$$

Every vector has an inverse (just flip all the signs). The ordinary real number zero will do as the zero vector. And of course, I can multiply any vector by a scalar:

$$7 \times (44x^5 + 19x^3 + 4x + 1) = 308x^5 + 133x^3 + 28x + 7$$

It all works. An obvious basis for this space would be the six vectors x^5, x^4, x^3, x^2, x, and 1. These vectors are linearly independent, because

$$ax^5 + bx^4 + cx^3 + fx^2 + gx + h = 0$$

(that is, the zero polynomial) for *any* value of the unknown x only if a, b, c, f, g, and h are all zero. And since any other vector in the space can be expressed in terms of these six, no group of seven or more vectors can be linearly independent. The space has dimension 6.

You might grumble that since I am not really doing any serious work on these polynomials—not, for example, trying to factorize them—the powers of x are really nothing more than place holders. What I am really playing with here is just the sextuplets of coefficients. Might I not just as well forget about x and write my three example polynomials as

$$(2, -8, -1, 11, -9, 15),$$
$$(44, 0, 19, 0, 4, 1),$$
$$(0, 0, 0, 1, -2, 1)?$$

If I then define addition of two sextuplets in the obvious way,

$$(a, b, c, d, e, f) + (p, q, r, s, t, u) = (a + p, b + q, c + r, d + s, e + t, f + u),$$

with scalar multiplication as

$$n \times (a, b, c, d, e, f) = (na, nb, nc, nd, ne, nf),$$

wouldn't this be, in some sense, the same vector space as the polynomial one?

Well, yes, it would. The mathematical object "vector space" is just a tool and an abstraction. We may choose to express it differently—geometrically, polynomially—to give us insights into the particular task we are using it for. But in fact *every* six-dimensional vector space over $\mathbb{R}$ is essentially the same as (mathematicians say "isomorphic to") the vector space of real-number sextuplets, with vector addition and scalar multiplication as I have defined them.

§VS.5   Vector spaces, while not very exciting in themselves, lead to much more potent and fascinating consequences when we (a) study their relations with each other or (b) enhance them slightly by adding new features to the basic model.

Under the first of those headings comes the large topic of *linear transformations*—the possible mappings of one vector space into another, assigning a vector in the second to a vector in the first, according to some definite rule. The word "linear" insists that these mappings be "well behaved"—so that, for example, if vector **u** maps into vector **f** and vector **v** maps into vector **g**, then vector **u** + **v** is guaranteed to map into vector **f** + **g**.

You can see the territory we get into here if you think of mapping a space of higher dimension into one of lower dimension—a "projection," you might say. Contrariwise, a space of lower dimension can be mapped into a space of higher dimension—an "embedding." We can also map a vector space into itself or into some lower-dimensional subspace of itself.

Since a number field like $\mathbb{Q}$ or $\mathbb{R}$ is just a vector space of dimension 1 over itself (you might want to pause for a moment to convince yourself of this), you can even map a vector space into its own scalar field. This type of mapping is called a *linear functional*. An example, using our vector space of polynomials, would be the mapping you get by just replacing x with some fixed value—say x equals 3—in every polynomial. This turns each polynomial into a number, linearly. Astoundingly (it has always seemed to me), the set of all linear functionals on a space forms a vector space by itself, with the functionals as vectors. This is the *dual* of the original space and has the same dimension. And why stop there? Why not map *pairs* of vectors into the ground field, so that any pair (**u**, **v**) goes to a scalar? This gets you the *inner product* (or *scalar product*) familiar to students of mechanics and quantum physics. We can even get more ambitious and map triplets, quadruplets, n-tuplets of vectors to a scalar, leading off into the theories of tensors, Grassmann algebras, and determinants. The humble vector space, though a simple thing in itself, unlocks a treasure cave of mathematical wonders.

§VS.6   So much for heading (a). What about (b)? What theories can
we develop by enhancing a vector space slightly, adding new features
to the basic model?

The most popular such extra feature is *some way to multiply vectors together.* Recall that in the basic definition of a vector space the
scalars form a field, with full addition, subtraction, multiplication,
and division. The vectors, however, can only be added and subtracted.
You can multiply them by scalars but not by each other. One obvious
way to enhance a vector space would be to add on some consistent
and useful method for multiplying two vectors together, the result
being another vector.

A vector space with this additional feature—that two vectors can
not only be added but also multiplied, giving another vector as the
result—is called *an algebra.*

This is, I agree, not a very happy usage. The word "algebra" already has a perfectly good meaning—the one that is the topic of this
book. Why confuse the issue by sticking an indefinite article in front
and using it to name this new kind of mathematical object? It's no
use complaining, though. The usage is now universal. If you hear
some mathematical object spoken of as "an algebra," it is almost certainly a vector space with some way to multiply vectors added on to it.

The simplest example of an algebra that is not completely trivial
is $\mathbb{C}$. Recall that the complex numbers can best be visualized by
spreading them out on an infinite flat plane, with the real part of the
number as an east-west coordinate, the imaginary part as a north-
south coordinate (Figure NP-4). This means that $\mathbb{C}$ can be thought of
as a two-dimensional vector space. A complex number is a vector; the
field of scalars is just $\mathbb{R}$; zero is the zero vector ($0 + 0i$, if you like); the
inverse of a complex number z is just its negative, $-z$; and the two
numbers 1 and i will serve as a perfectly good basis for the space. It's a
vector space . . . *with the additional feature* that two vectors—that is,
two complex numbers—can be multiplied together to give another
one. That means that $\mathbb{C}$ is not just a vector space, it is an algebra.

Turning a vector space into an algebra is not at all an easy thing to do, as the hero of my next chapter learned. That six-dimensional space of polynomial expressions I introduced a page or two ago, for instance, is not an algebra under ordinary multiplication of polynomials. This is why useful algebras tend to have names—there aren't that many of them. To make an algebra work at all, you may have to relax certain rules—the commutative rule in most cases. That's the rule that says that $a \times b = b \times a$. Often you have to relax the associative rule, too, the rule that says $a \times (b \times c) = (a \times b) \times c$.

If, as well as wanting to multiply vectors together, you also want to *divide* them, you have narrowed your options down even more dramatically. Unless you are willing to relax the associative rule, or allow your vector space to have an *infinite* number of dimensions, or allow your scalars to be something more exotic than ordinary numbers, you have narrowed those options all the way down to $\mathbb{R}$, $\mathbb{C}$, and the field of quaternions.

Ah, quaternions! Enter Sir William Rowan Hamilton.

Chapter 8

THE LEAP INTO
THE FOURTH DIMENSION

§8.1 MATHEMATICAL FICTION IS, TO PUT IT very mildly indeed, not a large or prominent category of literature. One of the very few such works to have shown lasting appeal has been Edwin A. Abbott's *Flatland*, first published in 1884 and still in print today.

Flatland is narrated by a creature who calls himself "A Square." He is, in fact, a square, living in a two-dimensional world—the Flatland of the book's title. Flatland is populated by various other living creatures, all having two-dimensional shapes of various degrees of regularity: triangles both isosceles (only two sides equal) and equilateral (all three sides equal), squares, pentagons, hexagons, and so on. There is a system of social ranks, creatures with more sides ranking higher, and circles highest of all. Women are mere line segments and are subject to various social disabilities and prejudices.

The first half of *Flatland* describes Flatland and its social arrangements. Much space is given over to the vexing matter of determining a stranger's social rank. Since a Flatlander's retina is one-dimensional (just as yours is two-dimensional), the objects in his field of vision are just line segments, and the actual shape of a stranger—and therefore his rank—is best determined by touch. A common form of introduction is therefore: "Let me ask you to feel Mr. So-and-so."

In the book's second half, A Square explores other worlds. In a dream he visits Lineland, a one-dimensional place, of which the author gives an 11-page description. Since a Linelander can never get past his neighbors on either side, propagation of the species presents difficult problems, which Abbott resolves with great delicacy and ingenuity.

A Square then awakens to his own world—that is, Flatland—where he is soon visited by a being from the third dimension: a sphere, who has a disturbing way of poking more or less of himself into Flatland, appearing to A Square as a circle mysteriously expanding and contracting. The sphere engages A Square in conversations of a philosophical kind, at one point introducing him to Pointland, a space of zero dimensions, inhabited by a being who "is himself his One and All, being really Nothing. Yet mark his perfect self-contentment, and hence learn this lesson, that to be self-contented is to be vile and ignorant, and that to aspire is to be blindly and ignorantly happy." I think we have all met this creature.

Abbott was not quite 46 years old when *Flatland* was published. The 240-word entry under his name in the 1911 *Encyclopædia Britannica* describes him as "English schoolmaster and theologian." It does not mention *Flatland*. Abbott was actually a boys school headmaster of a reforming and progressive cast of mind, inspired by a personal view of Christianity and skeptical of many of the conventions of Victorian society. *Flatland* was written secondarily as mild social satire.

In the 120-odd years since its publication, *Flatland* has caught the attention and stirred the imagination of countless readers. It has, in fact, generated a subgenre of spinoff books and stories. Dionys Burger's *Sphereland* and Ian Stewart's *Flatterland* are noteworthy elaborations on Abbott's original idea. The physics, chemistry, and physiology of two-dimensional existence—matters on which Abbott is light and not very convincing—were most brilliantly explored by A. K. Dewdney in his 1984 book *The Planiverse*.[75] Of lower literary quality, but sticking in the mind somehow, is Rudy Rucker's short story "Message Found in a Copy of *Flatland*," whose protagonist actually encounters Flatland in the basement of a Pakistani restaurant in London. He ends up eating the Flatlanders, who have "a taste something like very moist smoked salmon."[76]

Spaces of zero, one, two, and three dimensions—why stop there? Probably most nonmathematicians first heard about the fourth dimension from H. G. Wells's 1895 novel *The Time Machine*, whose protagonist says this:

> Space, as our mathematicians have it, is spoken of as having three dimensions, which one may call Length, Breadth, and Thickness, and is always definable by reference to three planes, each at right angles to the others. But some philosophical people have been asking why three dimensions particularly—why not another direction at right angles to the other three?—and have even tried to construct a Four-Dimensional geometry.

Now, it takes a while for abstruse theories to seep down into popular literature, even today.[77] If literary gents in the 1880s and 1890s were writing popular works about the number of dimensions space might or might not have, we can be sure that professional mathematicians must already have been pondering such questions among themselves for decades.

So they were. Ideas about the dimensionality of space were occurring to mathematicians here and there all through the second quarter of the 19th century. In the third quarter these occasional raindrops became a shower, allowing the great German mathematician Felix Klein (1849–1925) to make the following observation in hindsight: "Around 1870 the concept of a space of *n* dimensions became the general property of the advancing young generation [of mathematicians]."

Where did these ideas come from? There is no trace of them at the beginning of the 19th century. At the end of that century they are so widely known as to be showing up in popular fiction. Who first thought of them? Why did they appear at this particular time?

§8.2   During the early decades of the 19th century, mature ideas about complex numbers "settled in" as a normal part of the mental landscape of mathematicians. The modern conception of number, as I sketched it at the beginning of this book, was more or less established in mathematicians' minds (though the "hollow letters" notation I have used is late 20th century).

In particular, the usual pictorial representations of the real and complex numbers as, respectively, points spread out along a straight line, and points spread out over a flat plane, were common knowledge. The tremendous power of the complex numbers, their usefulness in solving a vast range of mathematical problems, was also widely appreciated. Once all this had been internalized, the following question naturally arose.

If passing from the real numbers, which are merely one-dimensional, to the complex numbers, which are two-dimensional, gives us such a huge increase in power and insight, why stop there? Might there not be other kinds of numbers waiting to be discovered— *hyper*-complex numbers, so to speak—whose natural representation is three-dimensional? And might those numbers not bring with them a vast new increase in our mathematical understanding?

This question had popped up in the minds of several mathematicians—including, inevitably, Gauss—since the last years of the 18th century, though without any very notable consequences. Around 1830, it occurred to William Rowan Hamilton.

§8.3   Hamilton's life[78] makes depressing reading. It is not that his circumstances were wretched in any outward way—marred by war, poverty, or sickness. Nor are there even any signs of real mental illness—chronic depression, for instance. Nor did he suffer from professional neglect or frustration—he was already famous in his teens. It is rather that Hamilton's life followed a downward trajectory. As a child he was a sensational prodigy; as a young man, merely a genius; in middle age, only brilliant; in his later years, a bore and a drunk.

Of Hamilton's mathematical talents there can be no doubt. He was possessed of great mathematical insight and worked very hard to bring that insight to bear on the most difficult problems of his day. Today mathematicians honor him and revere his memory.

Born in Dublin of Scottish parents, Hamilton is claimed by both Ireland and Scotland, but he spoke of himself as an Irishman. He was a child prodigy, accumulating languages until, at age 13, he claimed to have mastered one for each year of his age.[79] In the fall of 1823, he went up to Trinity College, Dublin, where he soon distinguished himself as a classics scholar. At the end of his first year, however, he met and fell in love with Catherine Disney.[80] Her family promptly married her off to someone more eligible, and this lost love blighted Hamilton's life and perhaps Catherine's too. Hamilton later (in 1833) married, more or less at random, a sickly and disorderly woman and suffered all his life from an ill-managed household.

Hamilton had picked up math in his teen years, quickly mastering the subject, and graduated from Trinity with the highest honors in both science and classics—an achievement previously unheard of. In his final year he came up with his "characteristic function," the ultimate source of the Hamiltonian operator so fundamental to modern quantum theory.

In 1827, Hamilton was appointed professor of astronomy at Trinity. Like all young mathematicians of his time, Hamilton was struck by the elegance and power of the complex numbers. In 1833, he produced a paper treating the system of complex numbers in a purely algebraic way, as what we should nowadays call "an algebra," in fact. For the complex number $a + bi$, Hamilton just wrote (a, b). The rule for multiplying complex numbers now becomes

$$(a, b) \times (c, d) = (ac - bd, ad + bc)$$

and the fact that i is the square root of minus one is expressed by

$$(0,\,1) \times (0,\,1) = (-1,\,0)$$

This may seem trivial, but that is because 170 years of mathematical sophistication stand between Hamilton and ourselves. In fact, it shifted thinking about complex numbers from the realms of arithmetic and analysis, where most mathematicians of the early 19th century mentally located them, into the realm of algebra. It was, in short, another move to higher levels of abstraction and generalization.

Hamilton now, from 1835 on, embarked on an eight-year quest to develop a similar algebra for bracketed *triplets*. Since passing from the one-dimensional real numbers to the two-dimensional complex numbers had brought such new riches to mathematics, what might not be achieved by advancing up one further dimension?

Getting a decent algebra out of number triplets proved to be very difficult, though. Of course, you can work *something* out if your conditions are loose enough. Hamilton knew, though, that for his triplets to be anything like as useful as complex numbers, their addition and multiplication needed to satisfy some rather strict conditions. They needed, for example, to obey a *distributive law*, so that if T_1, T_2, and T_3 are any triplets, the following thing is true:

$$T_1 \times (T_2 + T_3) = T_1 \times T_2 + T_1 \times T_3$$

They also needed, like the complex numbers, to satisfy a *law of moduli*. The modulus of a triplet $(a,\,b,\,c)$ is $\sqrt{a^2 + b^2 + c^2}$. The law says that if you multiply two triplets, the modulus of the answer is the product of the two moduli.

Hamilton fretted over this problem—the problem of turning his triplets into a workable algebra—more or less continuously through those eight years. During that time he acquired three children and a knighthood.

The mathematical conundrum was resolved for Hamilton with a flash of insight. He told the story himself, at the end of his life, in a letter to his second son, Archibald Henry:

Every morning in the early part of [October 1843], on my coming down to breakfast, your (then) little brother, William Edwin, and yourself, used to ask me, "Well, papa, can you multiply triplets?" Whereto I was always obliged to reply, with a sad shake of the head: "No, I can only add and subtract them."

But on the 16th day of the same month—which happened to be Monday, and a Council day of the Royal Irish Academy—I was walking in to attend and preside, and your mother was walking with me along the Royal Canal, to which she had perhaps driven; and although she talked with me now and then, yet an undercurrent of thought was going on in my mind which gave at last a result, whereof it is not too much to say that I felt at once the importance. An electric circuit seemed to close; and a spark flashed forth the herald (as I foresaw immediately) of many long years to come of definitely directed thought and work by myself, if spared, and, at all events, on the part of others if I should even be allowed to live long enough distinctly to communicate the discovery. Nor could I resist the impulse—unphilosophical as it may have been—to cut with a knife on a stone of Brougham Bridge, as we passed it, the fundamental formula with the symbols i, j, k:

$$i^2 = j^2 = k^2 = ijk = -1$$

which contains the Solution of the Problem, but, of course, the inscription has long since mouldered away.

The insight that came upon Hamilton that Monday by Brougham Bridge[81] was that triplets could not be made into a useful algebra, *but quadruplets could.* After the step from one-dimensional real numbers a to two-dimensional complex numbers $a + bi$, the next step was not to three-dimensional super-complex numbers $a + bi + cj$ but to *four*-dimensional ones $a + bi + cj + dk$. By dint of some simple rules for multiplying $i, j,$ and k, rules like those Hamilton inscribed on Brougham Bridge, these could be made into an algebra. Van der Waerden, in his *History of Algebra,* calls this inspiration "the leap into the fourth dimension."

§8.4   To make this new algebra work, Hamilton had to violate one
of the fundamental rules of arithmetic, the rule that says $a \times b = b \times a$.
I have already mentioned, in relation to Cauchy's compounding of
permutations (§7.7), that this is known as commutativity. The rule
$a \times b = b \times a$ is the *commutative rule*. It is familiar to all of us in
ordinary arithmetic. We know that multiplying 7 by 3 will give the
same result as multiplying 3 by 7. The commutative rule applies just
as well to complex numbers: Multiplying $2 - 5i$ by $-1 + 8i$ will give
the same result as multiplying $-1 + 8i$ by $2 - 5i$ ($38 + 21i$ in
both cases).

Quaternions can only be made to work as an algebra, though—
as a four-dimensional vector space with a useful way to multiply vectors—if you abandon this rule. In the realm of quaternions, for ex-
ample, $i \times j$ is not equal to $j \times i$; it is equal to its *negative*. In fact,

$$jk = i, \quad kj = -i$$
$$ki = j, \quad ik = -j$$
$$ij = k, \quad ji = -k$$

It was this breaking of the rules that made quaternions algebraically noteworthy and Hamilton's flash of insight one of the most
important revelations of that kind in mathematical history. Through
all the evolution of number systems, from the natural numbers and
fractions of the ancients, through irrational and negative numbers, to
the complex numbers and modular arithmetic that had exercised the
great minds of the 18th and early 19th centuries, the commutative
rule had always been taken for granted. Now here was a new system
of what could plausibly be regarded as numbers, in which the commutative rule no longer applied. The quaternions were, to employ
modern terminology, the first noncommutative division algebra.[82]

Every adult knows that when you have broken one rule, it is then
much easier to break others. As it is in everyday life, so it was with the
development of algebra. In fact, Hamilton described quaternions to a
mathematical friend, John Graves, in a letter dated October 17, 1843.
By December, Graves had discovered an *eight*-dimensional algebra, a

system of numbers later called *octonions*.[83] To make these work, though, Graves had to abandon yet another arithmetic rule, the *associative rule* for multiplication, the one that says $a \times (b \times c) = (a \times b) \times c$.

This, it should be remembered, was just the time when the non-Euclidean geometries—"curved spaces"—of Nikolai Lobachevsky and János Bolyai were becoming known. (I shall say more about this in §13.2.) Kant's notion that the familiar laws of arithmetic and geometry are inbuilt, immutable qualities of human thought were slipping away fast. If such fundamental arithmetic principles as the commutative and associative rules might be set aside, what else might? If four dimensions were needed to make quaternions work, who was to say that the world might not actually *be* four-dimensional?[84] Or that there might be creatures somewhere living in a two-dimensional Flatland?

§8.5   As brilliant as Hamilton's insight was, it was only one of a number of occurrences of four-dimensional thinking at around the same time. The fourth dimension, and the fifth, the sixth, and the *n*th dimensions too, were "in the air" in the 1840s. If the 1890s were the Mauve Decade, the 1840s were, at any rate among mathematicians, the Multidimensional Decade.

In that same year of 1843, in fact, the English algebraist Arthur Cayley, whom we shall meet again in the next chapter, published a paper titled "Chapters of Analytic Geometry of *n*-Dimensions." Cayley's intention was, as his title says, geometrical, but he used homogeneous coordinates (which I shall describe in my primer on algebraic geometry), and this gave the paper a strongly algebraic flavor.

In fact, homogeneous coordinates had been thought up by the German astronomer and mathematician August Möbius, who had published a classic book on the subject, *The Barycentric Calculus*, some years before. Möbius seems to have understood at that time that an irregular three-dimensional solid could be transformed into its mirror-image by a four-dimensional rotation. This—the date was

1827—may actually be the first trace of the fourth dimension in mathematical thinking.

The mention of Möbius here leads us to the Germans. Though these years were great ones for British algebraists, across the channel there was already swelling that tidal wave of talent that brought Germany to the front of the mathematical world and kept her there all through the later 19th and early 20th centuries.

§8.6   There was at that time a schoolmaster in the Prussian city of Stettin (now the Polish city of Szczecin) named Hermann Günther Grassmann. Thirty-four years old in 1843, he had been teaching in a *gymnasium* (high school, roughly) since 1831. He continued to teach school until nearly the end of his life. He was self-educated in mathematics; at university he had studied theology and philology. He married at age 40 and fathered 11 children.

The year after Hamilton's discovery, Grassmann published a book with a very long title beginning *Die lineale Ausdehnungslehre . . .* —"The Theory of Linear Extensions . . ." In this book—always referred to now as the *Ausdehnunsglehre* (pronounced "ows-DEHN-oongz-leh-reh")—Grassmann set out much of what became known, 80 years later, as the theory of vector spaces. He defined such concepts as linear dependence and independence, dimension, basis, subspace, and projection. He in fact went much further, working out ways to multiply vectors and express changes of basis, thus inventing the modern concept of "an algebra" in a much more general way than Hamilton with his quaternions. All this was done in a strongly algebraic style, emphasizing the entirely abstract nature of these new mathematical objects and introducing geometrical ideas as merely applications of them.

Unfortunately, Grassmann's book went almost completely unnoticed. There was just one review, written by Grassmann himself! Grassmann, in fact, belongs with Abel, Ruffini, and Galois in that sad company of mathematicians whose merit went largely unrecognized by their peers. In part this was his own fault. The *Ausdehnungslehre* is written in a style very difficult to follow and is larded with metaphysics in the early 19th-century manner. Möbius described it as "unreadable," though he tried to help Grassmann, and in 1847 wrote a commentary praising Grassmann's ideas. Grassmann did his best to promote the book, but he met with bad luck and neglect.

A French mathematician, Jean Claude Saint-Venant, produced a paper on vector spaces in 1845, the year after the *Ausdehnungslehre* appeared, showing ideas similar to some of Grassmann's, though plainly arrived at independently. After reading the paper, Grassmann sent relevant passages from the *Ausdehnungslehre* to Saint-Venant. Not knowing Saint-Venant's address, though, he sent them via Cauchy at the French Academy, asking Cauchy to forward them. Cauchy failed to do so, and six years later he published a paper that might very well have been derived from Grassmann's book. Grassmann complained to the Academy. A three-man committee was set up to determine whether plagiarism had occurred. One of the committee members was Cauchy himself. No determination was ever made. . . .

The *Ausdehnungslehre* did not go entirely unread. Hamilton himself read it, in 1852, and devoted a paragraph to Grassmann in the introduction to his own book, *Lectures on Quaternions*, published the following year. He praised the *Ausdehnungslehre* as "original and remarkable" but emphasized that his own approach was quite different from Grassmann's. Thus, nine years after the book's publication, precisely two serious mathematicians had noticed it: Möbius and Hamilton.

Grassmann tried again, rewriting the *Ausdehnungslehre* to make it more accessible and publishing a new edition of 300 copies in 1862 at his own expense. That edition contains the following preface, which I find rather moving:

> I remain completely confident that the labor I have expended on the science presented here and which has demanded a significant

part of my life as well as the most strenuous application of my powers, will not be lost. It is true that I am aware that the form which I have given the science is imperfect and must be imperfect. But I know and feel obliged to state (though I run the risk of seeming arrogant) that even if this work should again remain unused for another seventeen years or even longer, without entering into the actual development of science, still that time will come when it will be brought forth from the dust of oblivion and when ideas now dormant will bring forth fruit. I know that if I also fail to gather around me (as I have until now desired in vain) a circle of scholars, whom I could fructify with these ideas, and whom I could stimulate to develop and enrich them further, yet there will come a time when these ideas, perhaps in a new form, will arise anew and will enter into a living communication with contemporary developments. For truth is eternal and divine.

The 1862 edition fared little better than the 1844 one had, however. Disillusioned, Grassmann turned away from mathematics to his other passion, Sanskrit. He produced a massive translation into German of the Sanskrit classic *Rig Veda*, with a lengthy commentary—close to 3,000 pages altogether. For this work he received an honorary doctorate from the University of Tübingen.

The first major mathematical advance based directly on Grassmann's work came in 1878, the year after his death, when the English mathematician William Kingdon Clifford published a paper with the title "Applications of Grassmann's Extensive Algebra." Clifford used Grassmann's ideas to generalize Hamilton's quaternions into a whole family of n-dimensional algebras. These Clifford algebras proved to have applications in 20th-century theoretical physics. The modern theory of *spinors*—rotations in n-dimensional spaces—is descended from them.

§8.7   The 1840s thus brought forth two entirely new mathematical objects, even if they were not understood or named as such by their creators: the *vector space* and the *algebra*. Both ideas, even in their early primitive states, created wide new opportunities for mathematical investigation.

For practical application, too. These were the early years of the Electrical Age. At the time of Hamilton's great insight, Michael Faraday's discovery of electromagnetic induction was only 12 years in the past. Faraday himself was just 52 years old and still active. In 1845, two years AQ (After Quaternions), Faraday came up with the concept of an electromagnetic field. He saw it all in imaginative terms—"lines of force" and so on—having insufficient grasp of mathematics to make his ideas rigorous. His successors, notably James Clerk Maxwell, filled out the math and found vectors to be exactly what they needed to express these new understandings.

It is hard not to think, in fact, that interest in this wonderful new science of electricity, with currents of all magnitudes flowing in all directions, was one of the major impulses to vectorial thinking at this time.[85] Not that physicists found it easy to take vectors on board. There were three schools of thought, right up to the end of the 19th century and beyond.

The first school of vectorial thinking followed Hamilton, who had actually been the first to use the words "vector" and "scalar" in the modern sense. Hamilton regarded a quaternion $a + bi + cj + dk$ as consisting of a scalar part a and a vector part $bi + cj + dk$ and developed a way of handling vectors based on this system, with vectors and scalars rolled up together in quaternionic bundles.

A second school, established in the 1880s by the American Josiah Willard Gibbs and the Englishman Oliver Heaviside, separated out the scalar and vector components of the quaternion, treated them as independent entities, and founded modern vector analysis. The end result was an essentially Grassmannian system, though Gibbs testified that his ideas were already well formed before he picked up Grassmann's book, and Heaviside seems not to have read Grassmann at all. Gibbs and Heaviside were physicists, not mathematicians, and had the empirical attitude that the more snobbish kinds of pure mathematicians deplore. They just wanted some algebra that would work for them. If that meant taking a meat cleaver to Hamilton's quaternions, they had no compunctions about doing so.

A third school, exemplified by the British scientist Lord Kelvin (William Thomson), eschewed all this newfangled math completely and worked entirely in good old Cartesian coordinates x, y, and z. This cheerfully reactionary approach lingered for a long time, at any rate among the English, to whose rugged philistinism it had deep appeal. I learned dynamics in the 1960s from an elderly schoolmaster who was firmly in Lord Kelvin's camp and declared that vectors were "just a passing fad."

Disputes over the merits of these systems led to the slightly ludicrous Great Quaternionic War of the 1890s, of which there is a good account in Paul Nahin's 1988 biography of Oliver Heaviside (Chapter 9). Grassmann, or at any rate Gibbs/Heaviside, was the ultimate victor—the vector victor, if you like. Doing mathematical physics Lord Kelvin's way, with coordinates, came to seem quaint and cumbersome, while quaternions fell by the wayside.

Thus the algebra of quaternions never fulfilled Hamilton's great hopes for it, except indirectly. Instead of opening up broad new mathematical landscapes, as Hamilton had believed it would, and as he worked for the last 20 years of his life to ensure it would, the formal theory of quaternions turned out to be a mathematical backwater, of interest in a few esoteric areas of higher algebra but taught to undergraduates only as a brief sidebar to a course on group theory or matrix theory.[86]

§8.8   The study of n-dimensional spaces went off in other directions in the years AQ. In the early 1850s the Swiss mathematician Ludwig Schläfli worked out the geometry of "polytopes"—that is, "flat"-sided figures, the analogs of two-dimensional polygons and three-dimensional polyhedra—in spaces of four and more dimensions. Schläfli's papers on these topics, published in French and English in 1855 and 1858, went even more completely ignored than Grassmann's *Ausdehnungslehre* and only became known after Schläfli's death in 1895. This work properly belongs to geometry, though, not algebra.

Another line of development sprang from Bernhard Riemann's great "Habilitation" (a sort of second doctorate) lecture of 1854, titled *The Hypotheses That Underlie Geometry.* Picking up on some ideas Gauss had left lying around, Riemann shifted the entire perspective of the geometry of curves and surfaces, so that instead of seeing, say, a curved two-dimensional surface as being embedded in a flat three-dimensional space, he asked what might be learned about the surface from, so to speak, *within* it—by a creature unable to leave the surface. This "intrinsic geometry" generalized easily and obviously to any number of dimensions, leading to modern differential geometry, the calculus of tensors, and the general theory of relativity. Again, though, this is not properly an algebraic topic (though I shall return to it in §13.8, when discussing modern algebraic geometry).

§8.9   The theory of abstract vector spaces and algebras (vector spaces in which we are permitted to multiply vectors together in some manner) developed ultimately into the large area gathered today under the heading "Linear Algebra." Once you start liberating vector multiplication from the rigidities of commutativity and associativity, all sorts of odd things turn up and have to be incorporated into a general theory.

There are some algebras, for example, in which the zero vector has factors! Actually, Hamilton himself had noticed this when he tried to generalize his quaternions so that the coefficients a, b, c, and d in the quaternion $a + bi + cj + dk$ are not just real numbers (as he originally saw them) but complex numbers. Over the field of complex numbers, for instance, I can do this factorization:

$$x^2 + 1 = \left(x + \sqrt{-1}\right)\left(x - \sqrt{-1}\right)$$

(I have written " $\sqrt{-1}$ " there to avoid confusion with the i of Hamilton's quaternions, which is not quite the same thing as the i of complex numbers.) If I substitute Hamilton's j for x, I get

$$j^2 + 1 = \left(j + \sqrt{-1} \right)\left(j - \sqrt{-1} \right)$$

But by Hamilton's definition, $j^2 = -1$, so $j + \sqrt{-1}$ and $j - \sqrt{-1}$ are factors of zero. This is not a unique situation in modern algebra. Matrix multiplication, which I shall cover in the next chapter, will often give you a result matrix of zero when you multiply two nonzero matrices. This result does, though, show how quickly the study of abstract algebras slips away from the familiar world of real and complex numbers.

It is an interesting exercise to enumerate and classify all possible algebras. Your results will depend on what you are willing to allow. The narrowest case is that of commutative, associative, finite-dimensional algebras over (that is, having their scalars taken from) the field of real numbers $\mathbb{R}$ and with no divisors of zero. There are just two such algebras: $\mathbb{R}$ and $\mathbb{C}$, a thing proved by Karl Weierstrass in 1864. By successively relaxing rules, allowing different ground fields for your scalars, and permitting things like factors of zero, you can get more and more algebras, with more and more exotic properties. The American mathematician Benjamin Peirce carried out a famous classification along these lines in 1870.

A Scottish algebraist, Maclagen Wedderburn, in a famous 1908 paper titled "On Hypercomplex Numbers" took algebras to a further level of generalization, permitting scalars in any field at all . . . But now I have wandered into topics—fields, matrices—to which I have so far given no coverage. Matrices, in particular, need a chapter to themselves, one that starts out in Old China.

Chapter 9

An Oblong
Arrangement of Terms

§9.1   HERE IS A WORD PROBLEM. I found the problem easier to visualize if I thought of the three different kinds of grain as being different colors: red, blue, and green, for instance.

> *Problem.* There are three types of grain. Three baskets of the first, two of the second, and one of the third weigh 39 measures. Two baskets of the first, three of the second, and one of the third weigh 34 measures. And one basket of the first, two of the second, and three of the third weigh 26 measures. How many measures of grain are contained in one basket of each type?

Let's suppose that one basket of red grain contains x measures by weight, one basket of blue grain contains y measures, and one basket of green grain contains z measures. Then I have to solve the following system of simultaneous linear equations for x, y, and z:

$$3x + 2y + z = 39$$
$$2x + 3y + z = 34$$
$$x + 2y + 3z = 26$$

The solution, as you can easily check, is $x = \frac{37}{4}$, $y = \frac{17}{4}$, $z = \frac{11}{4}$.

Neither Diophantus nor the mathematicians of Old Babylon would have had much trouble with this problem. The reason it has such a prominent place in the history of mathematics is that the writer who posed it developed a systematic method for solving this and any similar problem, in any number of unknowns—a method that is still taught to beginning students of matrix algebra today. And all this took place over 2,000 years ago!

We do not know that mathematician's name. He was the author, or compiler, of a book titled *Nine Chapters on the Art of Calculation*. A collection of 246 problems in measurement and calculation, this was far and away the most influential work of ancient Chinese mathematics. Its precise influence on the development of medieval Indian, Persian, Muslim, and European mathematics is much debated, but versions of the book circulated all over East Asia from the early centuries CE onward, and given what we know of trade and intellectual contacts across Eurasia in the Middle Ages, it would be astonishing if some West-Asian and Western mathematics did not draw inspiration from it.

From internal evidence and some comments by the editors of later versions, we can place the original text of *Nine Chapters* in the former Han dynasty, which lasted from 202 BCE to 9 CE. This was one of the great epochs of Chinese history, the first in which the empire covered most of present-day metropolitan China,[87] and was securely unified under confident native rulers.

The Chinese culture area had actually been unified earlier by the famous and terrible "First Emperor" under his Qin dynasty in 221 BCE. After that tyrant died 11 years later, however, the Qin political system quickly fell apart. Years of civil war followed (providing China with a wealth of themes for literature, drama, and opera) before one of the warlords, a man named Liu Bang, obtained supremacy over his rivals and founded the Han dynasty in 202 BCE.

One of the Qin tyrant's most notorious deeds was the burning of the books. In accordance with the strict totalitarian doctrines of a philosopher named Shang Yang, Qin had ordered all books of speculative philosophy to be handed in to the authorities for burning. Fortunately, learning in ancient China was done mainly by rote memorization,[88] so after the Qin power collapsed, scholars with the destroyed texts still in their heads could reproduce them. Possibly this was the point when the *Nine Chapters* took decisive form, as a unified compilation of remembered texts from one or many sources. Or possibly not: The tyrant's edict exempted books on agriculture and other practical subjects, so that if the *Nine Chapters* existed earlier than the Han, it would not likely have been burned.

At any rate, the Early Han dynasty was a period of mathematical creativity in China. Peace brought trade, which demanded some computational skills. The standardization of weights and measures, begun by the Qin, led to an interest in the calculation of areas and volumes. The establishment of Confucianism as the foundation for state dogma required a reliable calendar so that the proper observances could be carried out at proper times. A calendar was duly produced, based on the usual 19-year cycle.[89]

Nine Chapters was probably one fruit of this spell of creativity. The book certainly existed by the 1st century CE and played a part in the subsequent mathematical culture of China comparable to that played by Euclid's *Elements* in Europe. And there, in the eighth chapter, is the grain measurement problem I described.

How does the author of *Nine Chapters* solve the problem? First, multiply the second of those equations by 3 (which will change it to $6x + 9y + 3z = 102$); then subtract the first equation from it *twice*. Similarly, multiply the third equation by 3 (making it $3x + 6y + 9z = 78$) and subtract the first equation from it *once*. The set of three equations has now been transformed into this:

$$3x + 2y + z = 39$$
$$5y + z = 24$$
$$4y + 8z = 39$$

Now multiply that third equation by 5 (making it $20y + 40z = 195$) and subtract the second equation from it four times. That third equation is thereby reduced to

$$36z = 99$$

from which it follows that $z = \frac{11}{4}$. Substituting this into the second equation gives the solution for y, and substituting the values of z and y into the first equation gives x.

This method is, as I said, a very general one, which can be applied not just to three equations in three unknowns but to four equations in four unknowns, five equations in five unknowns, and so on.

The method is known nowadays as Gaussian elimination. The great Carl Friedrich Gauss made some observations of the asteroid Pallas between 1803 and 1809 and calculated the object's orbit. This involved solving a system of six linear equations for six unknowns. Gauss tackled the problem just as I did above—which is to say, just as the unknown author of *Nine Chapters on the Art of Calculation* had 2,000 years previously.

§9.2   Once we have a good literal symbolism to hand, it is natural to wonder what solutions we would get for x, y, and z if we were to slog through the Gaussian elimination method for a *general* system of three equations, like this one:

$$ax + by + cz = e$$
$$fx + gy + hz = k$$
$$lx + my + nz = q$$

Here is the result:

$$x = \frac{bhq + ckm + egn - bkn - cgq - ehm}{agn + bhl + cfm - ahm - bfn - cgl}$$

$$y = \frac{ahq + ckl + efn - akn - cfq - ehl}{agn + bhl + cfm - ahm - bfn - cgl}$$

$$z = \frac{agq + bkl + efm - akm - bfq - egl}{agn + bhl + cfm - ahm - bfn - cgl}$$

That is the kind of thing that makes people give up math. If you persevere with it, though, you begin to spot some patterns among the spaghetti. The three denominators, for instance, are identical.

Let's concentrate on that denominator, the expression $agn + bhl + cfm - ahm - bfn - cgl$. Note that e, k, and q don't show up in it at all. It's constructed entirely from the coefficients at the left of the equals signs—that is, from this array:

$$a \quad b \quad c$$
$$f \quad g \quad h$$
$$l \quad m \quad n$$

Next thing to notice: None of the six terms in that expression for the denominator contain two numbers from the same row of the array or two from the same column.

Look at the ahm term, for instance. Having picked a from the first row, first column, it's as if the first row and the first column are now out of bounds. The next number, h, can't be taken from them; it has to be taken from elsewhere. And then, having taken h from the second row, third column, that row and column are then out of bounds, too, and there is no choice but to take m from the third row, second column.

It is not hard to show that applying this logic to a 3×3 array gives you six possible terms. For a 2×2 array, you would get two terms; for a 4×4 array, 24 terms. These are the factorial numbers I introduced in §7.4: $2! = 2$, $3! = 6$, $4! = 24$. The corresponding number of terms for five equations in five unknowns would be $5!$, which is 120.

The signs of the terms are more troublesome. Half are positive and half negative, but what determines this? Why is the agn term signed positive but ahm negative? Watch carefully.

First note that I was careful to write my terms, like that ahm, with the letters in order by the *row* I took them from: a from row 1, h from row 2, m from row 3. Then a term can be completely and unambiguously described by the *columns* the letters come from, in this case columns 1, 3, and 2. The quartet $(1, 3, 2)$ is a sort of alias for ahm, so long as I stick to my principle of writing the coefficients in row order, which I promise to do.

It is also, of course, a permutation of the basic triplet $(1, 2, 3)$, the permutation you get if you replace the basic 2 by 3 and the basic 3 by 2.

Now, permutations come in two varieties, odd and even. This particular permutation is odd, and that is why *ahm* has a minus sign in front. The term *bhl*, on the other hand, has the alias $(2, 3, 1)$, as you can easily check. This is the permutation you get to from basic $(1, 2, 3)$ if you replace 1 by 2, 2 by 3, and 3 by 1. It is an even permutation, and so *bhl* has a plus sign.

Wonderful, but how do you tell whether a permutation is odd or even? Here's how. I'll continue using the expression for three equations with three unknowns, but all of this extends easily and obviously to 4, 5, or any other number of equations.

Form the product $(3 - 2) \times (3 - 1) \times (2 - 1)$ using every possible pair of numbers $(A - B)$ for which $A > B$ and A and B are either 1, 2, or 3. The value of this product is 2 (which is $2! \times 1!$, so you can easily see the generalization. If we were working with four equations in four unknowns, the product would be $(4 - 3) \times (4 - 2) \times (4 - 1) \times (3 - 2) \times (3 - 1) \times (2 - 1)$, which is 12, which is $3! \times 2! \times 1!$). This value, however, doesn't actually matter. What matters is its *sign*, which of course is positive.

Now just walk through that expression applying some permutation to the numbers 1, 2, and 3. Try that first permutation, the one for *ahm*, where we replace the basic 2 by 3 and 3 by 2. Now the expression looks like $(2 - 3) \times (2 - 1) \times (3 - 1)$, which works out to –2. So this is an odd permutation. Applying the *bhl* permutation, on the other hand, changes $(3 - 2) \times (3 - 1) \times (2 - 1)$ into $(1 - 3) \times (1 - 2) \times (3 - 2)$, which is +2. This is an even permutation.

This business of even and odd permutations is an important one, obviously related to the issues I discussed in §§7.3-7.4. Here are all six possible permutations of $(1, 2, 3)$, with their parity (even or odd) worked out by the method I just used:

$$(1, 2, 3) \quad (3 - 2) \times (3 - 1) \times (2 - 1) = 2$$
$$(2, 3, 1) \quad (1 - 3) \times (1 - 2) \times (3 - 2) = 2$$
$$(3, 1, 2) \quad (2 - 1) \times (2 - 3) \times (1 - 3) = 2$$
$$(1, 3, 2) \quad (2 - 3) \times (2 - 1) \times (3 - 1) = -2$$
$$(3, 2, 1) \quad (1 - 2) \times (1 - 3) \times (2 - 3) = -2$$
$$(2, 1, 3) \quad (3 - 1) \times (3 - 2) \times (1 - 2) = -2$$

As you can see, half have even parity, half odd. It always works out like this.

So the sign on each term in that denominator expression is determined by the sign of the alias permutation that corresponds to the coefficients. Whew!

The denominator expression I have been working over here is an example of a *determinant*. So, as a matter of fact, are the numerators I got for x, y, and z. You might try figuring out which 3×3 array each numerator is the determinant of. The study of determinants led eventually to the discovery of *matrices*, now a vastly important subtopic within algebra. A matrix is an array of (usually) numbers, like that 3×3 one I set out above but treated *as an object in its own right*. More on this shortly.

It is an odd thing that while a modern course in algebra introduces matrices first and determinants later, the historical order of discovery was the opposite: determinants were known long before matrices.

The fundamental reason for this, which I shall enlarge on as I go along, is that a determinant is a *number*. A matrix is not a number. It is a different kind of thing—a different *mathematical object*. And the period we have reached in this book (though not yet in this chapter, which still has some prior history to fill in) is the early and middle 19th century, when algebraic objects were detaching themselves from the traditional mathematics of number and position and taking on lives of their own.

§9.3   Although mathematicians doodling with systems of simultaneous linear equations must have stumbled on determinant-type expressions many times over the centuries, and some of the algebraists I
have already mentioned—notably Cardano and Descartes—came
close to discovering the real thing, the actual, clear, indisputable discovery of determinants did not occur until 1683. It is one of the most
remarkable coincidences in the history of mathematics that the discovery of determinants took place *twice* in that year. One of these
discoveries occurred in the kingdom of Hanover, now part of Germany; the other was in Edo, now known as Tokyo, Japan.

The German mathematician is of course the one more familiar
to us. This was the great Gottfried Wilhelm von Leibniz, co-inventor
(with Newton) of the calculus, also philosopher, logician, cosmologist, engineer, legal and religious reformer, and general polymath—
"a citizen of the whole world of intellectual inquiry," as one of his
biographers says.[90] After some travels in his youth, Leibniz spent the
last 40 of his 70 years in service to the Dukes of Hanover, one of the
largest of the petty states that then occupied the map of what is now
Germany.

In a letter to the French mathematician-aristocrat the Marquis
de l'Hôpital, written in that year of 1683, Leibniz said that if this
system of simultaneous linear equations—*three* equations in *two* unknowns, note—

$$a + bx + cy = 0$$
$$f + gx + hy = 0$$
$$l + mx + ny = 0$$

has solutions x and y, then

$$agn + bhl + cfm = ahm + bfn + cgl$$

This is the same as saying that the determinant $agn + bhl + cfm -
ahm - bfn - cgl$ must be zero. Leibniz was quite right, and although he
did not construct a full theory of determinants, he did understand
their importance in solving systems of simultaneous linear equations

and grasped some of the symmetry principles that govern their structure and behavior, principles like those I sketched above.

Leibniz's Japanese co-discoverer, of whom he lived and died perfectly unaware, was Takakazu Seki, who was born in either 1642 or 1644 in either Huzioka or Edo. "Knowledge of Seki's life is meager and indirect," says Akira Kobori in the *DSB* entry for Seki. His biological father was a samurai, but the Seki family adopted him, and he took their surname.

Japan was at this time a few decades into an era of national unity and confidence under strong rulers—the Edo period, one of whose first and greatest Shoguns was the subject of a colorful novel by James Clavell. Unification and peace had allowed a money economy to develop, so that accountants and comptrollers were in great demand. The patriarch of the Sekis was in this line of work, and Takakazu followed in his footsteps, rising to become chief of the nation's Bureau of Supply and being promoted to the ranks of samurai himself. "In 1706 [I am quoting from the *DSB* again] having grown too old to fulfill the duties of his office, he was transferred to a sinecure and died two years later."

The first math book ever written by a Japanese had appeared in 1622, Sigeyosi Mori's *A Book on Division*. Our man Seki was a grand-student of Mori's—I mean, Seki's teacher (a man named Takahara, about whom we know next to nothing) was one of Mori's students. Seki was also strongly influenced by Chinese mathematical texts—no doubt he knew the *Nine Chapters on the Art of Calculation*.

Seki went far beyond the methods known in East Asia at that time, though, developing a "literal" (the "letters" were actually Chinese characters) symbolism for coefficients, unknowns, and powers of unknowns. Though his solutions for equations were numerical, not strictly algebraic, his investigations went deep, and he came within a whisker of inventing the calculus. What we nowadays call the Bernoulli numbers, introduced to European math by Jacob Bernoulli in 1713, had actually been discovered by Seki 30 years earlier.[91]

As a samurai, Seki was expected to practice modesty, and that apparently precluded his publishing books in his own name. There was also a culture of secrecy between rival schools of mathematical instruction, rather like the one in 16th-century Italy that I have already described. What we know of Seki's work is taken from two books published by his students, one in 1674 and one posthumously in 1709. It is in this second book that Seki's work on determinants appears. He picked up and generalized the Chinese method of elimination by rows, the method that I described earlier, and showed the part played by the determinant in it.

§9.4   Unfortunately, neither Seki's work nor that of Leibniz had much immediate consequence. There seems to have been no further development in Japan at all until the modern period. In Europe an entire lifetime passed before determinants were taken up again. Then, suddenly, they were "in the air" and by the late 1700s had entered the Western mathematical mainstream.

The process of solving a system of simultaneous linear equations by use of determinants, the process I sketched out in §9.2, is known to mathematicians as Cramer's rule. It first appeared in a book titled *Introduction to the Analysis of Algebraic Curves*, published in 1750. The author of the book, Gabriel Cramer, was a Swiss mathematician and engineer, widely traveled and well acquainted with all the great European mathematicians of his day. In his book he tackles the problem of finding the simplest algebraic curve (that is, a curve whose x, y equation has a polynomial on the left of the equals sign and a zero on the right) passing through n arbitrary points in a flat plane. He found that, given an arbitrary five points, we can find a second-degree curve to fit them—a curve, that is, with an equation like this:

$$ax^2 + 2hxy + by^2 + 2gx + 2fy + c = 0$$

I shall have much more to say about this kind of thing later, in my primer on algebraic geometry. Finding the equation of the actual curve for a given actual five points leads to a system of five simultaneous linear equations in five unknowns. This is not merely an abstract problem. By Kepler's laws, planets move on curves of the second degree (to a good approximation anyway), so that five observations of a planet's position suffice to determine its orbit fairly accurately.[92]

§9.5   Can you make determinants interact with each other? Given two square arrays, for instance, if I were to add their corresponding elements to get a new array, like

$$\begin{matrix} a & b \\ c & d \end{matrix} + \begin{matrix} p & q \\ r & s \end{matrix} = \begin{matrix} a+p & b+q \\ c+r & d+s \end{matrix},$$

is the determinant of that new array the sum of the determinants of the first two? Alas, no: The determinants of the two arrays on the left sum to $(ad-bc)+(ps-qr)$, while the determinant of the array on the right is $(a+p)\times(d+s)-(b+q)\times(c+r)$. The equality is not true.

While adding determinants doesn't get you anywhere much, *multiplying* them does. Let me just multiply those two determinants: $(ad-bc)\times(ps-qr)$ is equal to $adps+bcqr-adqr-bcps$. Is that the determinant of any interesting array? As a matter of fact it is. It is the determinant of this 2×2 array:

$$\begin{matrix} ap+br & aq+bs \\ cp+dr & cq+ds \end{matrix}$$

If you stare hard at that array, you will see that every one of its four elements is gotten by simple arithmetic on a *row* from the first array—either a, b or c, d—and a *column* from the second—either p, r or q, s. The element in the second row, first column, for example, comes from the second row of the first array and the first column of the second. This doesn't just work for 2×2 arrays either: If you multiply the determinants of two 3×3 arrays, you get an expression which is the determinant of another 3×3 array, and the number in the nth row, mth column of this "product array" is gotten by merging the nth row of the first array with the mth column of the second, in a procedure just like the one described above.[93]

To a mathematician of the early 19th century, looking at that 2×2 product array would bring something else to mind: §159 of Gauss's great 1801 classic *Disquisitiones arithmeticae*. Here Gauss asks the following question: Suppose, in some expression for x and y, I make the substitutions $x = au + bv$, $y = cu + dv$. In other words, I am changing my unknowns from x and y to u and v by a *linear transformation—x* a linear expression in u and v, and y likewise. And suppose then I do this substitution business *again*, switching to yet another couple of unknowns w and z via two more linear transformations: $u = pw + qz$, $v = rw + sz$.

What's the net effect? In going from unknowns x and y to unknowns w and z via those intermediate unknowns u and v, using linear transformations all along the way, what is the substitution I end up with? It's easy to work out. The net effect is this substitution: $x = (ap + br)w + (aq + bs)z$, $y = (cp + dr)w + (cq + ds)z$. Look at the expressions in the parentheses! It seems that multiplying determinants might have something to do with linear transformations.

With these ideas and results floating around, it was only a matter of time before someone established a good coherent theory of determinants. This was done by Cauchy, in a long paper he read to the French Institute in 1812. Cauchy gave a full and systematic description of determinants, their symmetries, and the rules for manipulating them. He also described the multiplication rule I have given here, though of course in much more generality than I have given. Cauchy's 1812 paper is generally considered the starting point of modern matrix algebra.

§9.6   It took 46 years to get from the manipulation of determinants to a true abstract algebra of matrices. For all the intriguing symmetries and rules of manipulation, determinants are still firmly attached to the world of numbers. A determinant *is* a number, though one whose calculation requires us to go through a complicated algebraic expression. A matrix in the modern understanding is not a number. It is an *array*, like the ones I have been dealing with. The elements stacked in its rows and columns might be numbers (though this is not necessarily the case), and there will be important numbers associated with it—notably its determinant. The matrix, however, is a thing of interest to mathematicians in itself. It is, in short, a new mathematical object.

We can add or subtract two square[94] matrices to get another one; we just add the arrays term by term. (This works out to be a suitable procedure for matrices, even though the associated determinants come out wrong.) We can multiply a matrix by an ordinary number to get a different matrix. Does this sound familiar? The family of all $n \times n$ matrices forms a vector space, of dimension n^2. More than that: We can, using the techniques I illustrated above for determinants, multiply matrices together in a consistent way, so the family of all $n \times n$ matrices forms not merely a vector space but an algebra!

We can make a case that this is the most important of all algebras. For example, it encompasses many other algebras. The algebra of complex numbers, to take a simple case, can be matched off precisely—"mapped," as we say—with all its rules of addition, subtraction, multiplication, and division, to a certain subset of the 2×2 matrices. You might care to confirm that the rule for matrix multiplication (it is the same as the one for determinants that I gave in the preceding section) does indeed reproduce complex-number multiplication when complex numbers $a + bi$ and $c + di$ are suitably represented by matrix equivalents:

$$\begin{pmatrix} a & b \\ -b & a \end{pmatrix} \times \begin{pmatrix} c & d \\ -d & c \end{pmatrix} = \begin{pmatrix} (ac - bd) & (ad + bc) \\ -(ad + bc) & (ac - bd) \end{pmatrix}$$

You might want to figure out which matrix represents i in this scheme of things. Then multiply that matrix by itself and confirm that you do indeed get the matrix representing -1.

Hamilton's quaternions can be similarly mapped into a family of 4×4 matrices. The fact of their multiplication not being commutative doesn't matter, since matrix multiplication is not commutative either (though a particular family of matrices, like the one representing complex numbers, might preserve commutativity within itself). Noncommutativity was cropping up all over mid-19th-century algebra. Permutations, also noncommutative, can be represented by matrices, too, though the math here would take us too far afield.

Matrices are, in short, the bee's knees. They are tremendously useful, and any modern algebra course quite rightly begins with a good comprehensive introduction to matrices.

§9.7   I said that it took 46 years to get from determinants to matrices. Cauchy's definitive paper on determinants was read to the French Institute in 1812. The first person to use the word "matrix" in this algebraic context was the English mathematician J. J. Sylvester in a scholarly article published in 1850. Sylvester defined a matrix as "an oblong arrangement of terms." However, his thinking was still rooted in determinants. The first formal recognition of a matrix as a mathematical object in its own right came in another paper, "Memoir on the Theory of Matrices," published in the *Transactions of the London Philosophical Society* by the English mathematician Arthur Cayley in 1858.

Cayley and Sylvester are generally covered together in histories of mathematics, and I see no reason to depart from this tradition. They were near coevals, Sylvester (born 1814) seven years the older. They met in 1850, when both were practicing law in London, and became close friends. Both worked on matrices; both worked on invariants (of which more later). Both studied at Cambridge, though at different colleges.

Cayley was elected a fellow of his college—Trinity, Cambridge—and taught there for four years. To continue, he would have had to take holy orders in the Church of England, which he was not willing to do. He therefore went into the law, being admitted to the bar in 1849.

Sylvester's first job was as a professor of natural philosophy at the new University of London. De Morgan (see §10.3) was one of his colleagues. In 1841, however, Sylvester left to take up a professorship at the University of Virginia. That lasted three months; then he resigned after an incident with a student. The incident is variously reported, and I don't know which report is true. It is clear that the student insulted Sylvester, but the stories differ about what happened next. Sylvester struck the student with a sword-stick and refused to apologize; or Sylvester demanded that the university discipline the student, but the university would not. There is even a theory involving a homoerotic attachment. Sylvester never married, wrote florid poetry, enjoyed singing in a high register, and brought forth the following comment from diarist and mathematical hanger-on Thomas Hirst:

> On Monday having received a letter from Sylvester I went to see him at the Athenaeum Club . . . He was, moreover, excessively friendly, wished we lived together, asked me to go live with him at Woolwich and so forth. In short he was eccentrically affectionate.

Whatever the facts of the case, I don't think we need to venture into speculations about Sylvester's inner life to see that the traits of character noted above, together with his Jewishness (he was born with the surname Joseph; "Sylvester" was a later family addition) and his anti-slavery opinions, would not have done much to commend him to the young bloods of antebellum Charlottesville. He returned to England, took a job as an actuary, studied for the bar, and supplemented his income by taking private students (one of whom was Florence Nightingale, "the lady with the lamp," who was a very capable mathematician and statistician).

Cayley and Sylvester were just two of a fine crop of algebraists that came up in the British Isles in the early 19th century. Hamilton, of course, belongs in this company, too. It is, in fact, worth spending another chapter in what one of Thackeray's characters calls "these brumous isles" to better understand some of the background to the great transformation of algebraic thinking in the early and middle years of that century.

Chapter 10

Victoria's Brumous Isles

§10.1   HERE IS ENGLISH MATHEMATICIAN George Peacock in his *Treatise on Algebra*, published in 1830:

> [Arithmetic] can only be considered as a Science of Suggestion, to which the principles and operations of Algebra are adapted, *but by which they are neither limited nor determined.*

(My italics.) Here, 10 years later, is the young (he was 27 at the time) Scottish mathematician Duncan Gregory, who had studied under Peacock:

> There are a number of theorems in ordinary algebra, which, though apparently proved to be true only for symbols representing numbers, admit of a much more extended application. Such theorems depend only on the laws of combination to which the symbols are subject, and are therefore true for all symbols, whatever their nature may be, which are subject to the same laws of combination.

And here is another Englishman, Augustus De Morgan, in his *Trigonometry and Double Algebra* (1849):

Given symbols M, N, +, and one sole relation of combination, namely that $M + N$ is the same as $N + M$. Here is a symbolic calculus: how can it be made a significant one? In the following ways, among others. 1. M and N may be magnitudes, and + the sign of addition of the second to the first. 2. M and N may be numbers, and + the sign of multiplying the first by the second. 3. M and N may be lines, and + the direction to make a rectangle with the antecedent for a base, and the consequent for an altitude. 4. M and N may be men and + the assertion that the antecedent is the brother of the consequent. 5. M and N may be nations, and + the sign of the consequent having fought a battle with the antecedent.

Plainly algebra was cutting loose from the world of numbers in the second quarter of the 19th century. What was driving this process? And why were those declarations all uttered by mathematicians from the British Isles?

§10.2   As the 18th century progressed, British mathematics lagged further and further behind developments on the continent. In part this was Sir Isaac Newton's fault; or rather, it was a by-product of the swelling self-regard of the British, for most of whom Sir Isaac was a national hero. This swelling action had, in the proper Newtonian manner, an equal and opposite reaction: The great continental nations set up their own culture heroes in opposition to Newton. Descartes served this purpose for the French. The aforementioned book by Patricia Fara[95] records a patriotic British drinking song from around 1760:

> The atoms of [Des]Cartes Sir Isaac destroyed
> Leibnitz [*sic*] pilfered our countryman's fluxions;
> Newton found out attraction, and prov'd nature's void
> Spite of prejudiced Plenum's constructions.
> Gravitation can boast,
> In the form of my toast,
> More power than all of them knew, Sir.

The Germans in fact had two anti-Newton icons: not only Leibniz but also Goethe, who was a bitter critic of Newton's optical theories. "Goethe's house in Weimar is still decorated with a defiantly anti-Newtonian rainbow," Ms. Fara tells us.[96]

All this was unfortunate for British mathematics because the notation Newton had devised for the operations of the calculus was definitely inferior to the one promoted by Leibniz and then taken up all over the continent. Patriotic Britons stuck with Newton's "dot" notation instead of taking up Leibniz's d's—that is, writing, for example, $\ddot{x}$ where the continentals wrote

$$\frac{d^2 x}{dt^2}$$

This had an isolating and retarding effect on British calculus.[97] It made British papers tiresome for continental mathematicians to read and obscured the fact, whose significance was now dawning, that x (here a function of t) is being acted on by an operator,

$$\frac{d^2}{dt^2},$$

that could be detached and considered a mathematical object in its own right.

Even allowing for the Newton factor, though, it is hard to avoid the impression that stiff-necked national pride and insularity were independently working to hold British mathematics back. Complex numbers, for example, had long since "settled in" to European mathematics. In Britain, by contrast, even negative numbers were still scorned by some professional mathematicians, as witness the following, taken from the preface to William Frend's *Principles of Algebra* (1796):

> [A number] submits to be taken away from a number greater than itself, but to attempt to take it away from a number less than itself is ridiculous. Yet this is attempted by algebraists who talk of a number less than nothing; of multiplying a negative number into a negative

number and thus producing a positive number; of a number being imaginary. . . . This is all jargon, at which common sense recoils; but from its having been once adopted, like many other figments, it finds the most strenuous supporters among those who love to take things upon trust and hate the colour of a serious thought.

Frend was no lone crank, either. He had been Second Wrangler—that is, second in the year's mathematics examination—at Christ's College, Cambridge, in 1780. He became one of Britain's first actuaries. Later he struck up a friendship with that Augustus De Morgan I quoted from in the previous section, and De Morgan married one of his daughters.

By the early years of the 19th century, the younger generation of British mathematicians had become dissatisfied with this state of affairs. The long wars against Napoleon had had the double effect of forcing Britons to pay attention to continental ideas more than formerly and of bringing home to mathematicians of the offshore nation (a single United Kingdom since the 1801 Act of Union) how very good French mathematics was.

In 1813, three young scholars at Trinity College, Cambridge—Sir Isaac Newton's old college—took action, founding what they called the Analytical Society. These three scholars, all born in 1791 or 1792, were John Herschel, son of the astronomer who had discovered Uranus; Charles Babbage, later famous for his "calculating engine" (a sort of mechanical computer); and the George Peacock I quoted above. The main purpose of their society was to reform the teaching of calculus, promoting, as Babbage punned, "the principles of pure d-ism as opposed to the *dot*-age of the university."

The Analytical Society does not seem to have lasted very long, and none of the three founders attained the first rank in mathematics, but the spirit the society embodied was carried forward by Peacock, an energetic and idealistic reformer, very much in the style of his time. After graduating he became a lecturer at Trinity and then, in 1817, an examiner in mathematics. His first act on being appointed examiner was to switch the calculus teaching from Newton's *dot*-age to Leibniz's *d*-ism.

Peacock went on to become a full professor and was instrumental in the establishment of several learned societies, notably the Astronomical Society of London, the Philosophical Society of Cambridge, and the British Association for the Advancement of Science. All these new societies were open to any person of ability and accomplishment, a break with the older idea of a learned society as a sort of gentlemen's club that carefully excluded self-educated working people and "rude mechanicals." The technocratic lower-middle classes of the early Industrial Revolution were flexing their muscles. Peacock ended his days happily as dean of Ely cathedral in eastern England.

This general spirit of reform is the background to British mathematics in this period. Its fruits can be seen in the next generation of British mathematicians, most especially algebraists. I have already described the work of Hamilton, born in 1805. Close behind came De Morgan (born 1806), J. J. Sylvester (1814), George Boole (1815), and Arthur Cayley (1821). These men rescued their country's mathematical reputation, at least in algebra. To them we owe all or part of the theory of groups, the theory of matrices, the theory of invariants, and the modern theory of the foundations of mathematics.[98]

§10.3   Augustus De Morgan is the least mathematically consequential of the four men I just named but in many ways the most interesting. He also has a special place in this author's heart, having served as the very first professor of mathematics at my own alma mater: University College, London—"the godless institution on Gower Street."

The great old English universities of Oxford and Cambridge entered the 19th century still cumbered with much religious, social, and political baggage left over from their earlier histories. Neither accepted women, for example; both required a religious test—basically, a declaration of loyalty to the Church of England and its doctrines— for masters and fellows. (Oxford actually required it for graduation.)

These restrictions[99] had come to seem absurd to a great many people, and once the Napoleonic Wars were done with and a spirit of reform was in the air, there was a rising sentiment in the British intellectual classes for a more progressive institution of higher education. This sentiment found practical expression at last in the founding of the University of London, which admitted its first students in 1828.

The new university was the first institution of higher education in England to accept students of any sex, religion, or political opinion.[100] Other colleges quickly came up in other parts of the city. The University of London is now, in the early 21st century, multicollegiate, like the older universities, and the original Gower Street establishment is known as University College.

The founding of this new university was very timely for Augustus De Morgan. He had taken his bachelor's degree at Trinity College, Cambridge. Peacock had been one of his teachers; another had been George Airy, who later became Astronomer Royal, and who has a mathematical function named after him.[101] On graduating fourth in the mathematical exams in 1826, De Morgan contemplated taking a master's degree. However, that religious test was required. De Morgan seems to have been of a naturally ("deeply" says his biographer W. S. Jevons) religious disposition, but his religion was personal, and he was no friend of any organized church, certainly not the Church of England.

Always a man of strong principle, De Morgan declined to take the tests, went home to London, and, like Cayley 20 years later, resigned himself to becoming a lawyer. He had barely registered at Lincoln's Inn, however, when the new university opened, and he was offered the chair of mathematics. He took it, delivering his first lecture—"On the Study of Mathematics"—in Gower Street at the age of 22. De Morgan then held the professorship until 1866, when he resigned on a point of principle.

A bookish and good-natured man, De Morgan strikes the reader of his biographies as a person one would like to have invited to dinner. A great popularizer of science, he contributed eagerly to the many societies and little magazines that catered to the rising technical and commercial classes of late-Georgian, early-Victorian Britain. ("His articles of various length cannot be less in number than 850"— Jevons.) He was a bibliophile and a good amateur flautist. His wife ran an intellectual salon, in the old French style, from their home at 30 Cheyne Walk, Chelsea. His daughter wrote fairy tales. A particularly creepy one, "The Hair Tree," haunted my own childhood.

De Morgan had a puzzler's mind, with a great love of verbal and mathematical curiosities, some of which he collected in his popular book *A Budget of Paradoxes* (printed posthumously by his widow in 1872). He was especially pleased to know that he was x years old in the year x^2, a distinction that comes to very few,[102] and that his name was an anagram of: "O Gus! Tug a mean surd!"

§10.4   De Morgan's importance for the history of algebra is his attempt to overhaul logic and improve its notations. Logic had undergone very little development since its origins under Aristotle. As taught up to De Morgan's time, it rested on the idea that there were four fundamental types of propositions, two affirmative and two negative. The four types were:

Universal, affirmative ("All X is Y")
Particular, affirmative ("Some X is Y")
Universal, negative ("No X is Y")
Particular, negative ("Some X is not Y")

Such propositions can be combined in sets of three, called syllogisms, two premises leading to a conclusion:

All men are mortal
Socrates is a man
_____

Socrates is mortal

There was in De Morgan's time a Sir William Hamilton, professor of logic and metaphysics at Edinburgh, *not* the same person as the Sir William Rowan Hamilton of my Chapter 8, though the two are often confused.[103] In some 1833 lectures on logic, this "other Hamilton" suggested an improvement to the Aristotelian scheme. He thought it wrong of Aristotle to have quantified the subjects of his propositions ("All X . . . ," "Some X . . .") but not the predicates (". . . is Y," . . . is not Y"). His suggested improvement was the *quantification of the predicate.*

De Morgan took up this idea and ran with it, eventually producing a book titled *Formal Logic, or the Calculus of Inference, Necessary and Probable* (1847). He followed the book in subsequent years with four further memoirs on the subject, intending all these writings to stand as a vast new system of logic built around an improved notation and the quantification of the predicate. Sir William Hamilton, who had supplied the original idea, was not much impressed. He referred to De Morgan's system as "horrent with spiculae" (that is, bristling with spikes). It is nowadays only a historical curiosity, since De Morgan merely improved the traditional way of writing out logical formulas. What was really needed for progress in logic was a fully modern algebraic symbolism. That was supplied by George Boole.

§10.5   Boole was one of the "new men" of early 19th-century Britain, from humble origins and self-taught, financed by no patron and with nothing but his own merit and energy to help him rise. The son of a small-town cobbler and a lady's maid, Boole got such learning as his parents could afford, supplementing it with intensive studies of his own. At age 14, he was producing translations of Greek verse. When George was 16, however, his father's affairs collapsed, and George had to take a job as a schoolmaster to support the family. He continued schoolmastering for 18 years, running his own schools for the most part. He opened his first when he was just 19.

— Before Universal Arithmetic —

Otto Neugebauer (1899–1990) found algebra in old Babylonian tablets.

The last moments of Hypatia (c.370–415), in the Victorian imagination.

Omar Khayyam (1048–1131) wrote poetry and tackled the cubic equation.

Girolamo Cardano (1501–1576) found a general solution for the cubic.

— Using Letters for Numbers —

François Viète (1540–1603) separated
"things sought" from "things given."

René Descartes (1596–1650)
algebraized geometry.

Sir Isaac Newton (1642–1727)
saw symmetry in solutions.

Gottfried von Leibniz (1646–1716)
found relief for his imagination.

— From Equations to Groups —

Joseph-Louis Lagrange (1736–1813)
carried symmetry forward.

Paolo Ruffini (1765–1822) believed
the quintic was unsolvable.

Augustin-Louis Cauchy (1789–1857)
made an "arithmetic" of
permutations.

Niels Abel (1802–1829)
proved Ruffini correct

— Discovery of the Group —

Évariste Galois (1811–1832) found permutation groups in equations.

Arthur Cayley (1821–1895) abstracted the group idea.

Ludwig Sylow (1832–1918) delved into the structure of finite groups.

Camille Jordan (1838–1922) wrote the first book on groups.

— Into the Fourth Dimension —

*Sir William R. Hamilton (1805–1865)
found a new algebra.*

*Hermann Grassmann (1809–1877)
explored vector spaces.*

*Bernhard Riemann (1826–1866)
launched two geometric revolutions.*

*Edwin A. Abbott (1838–1926) took us
to Flatland.*

Julius Plücker (1801–1868) based his geometry on lines, not points.

Sophus Lie (1842–1899) mastered continuous groups.

Felix Klein (1849–1925) urged the group-ification of geometry.

Henri Poincaré (1854–1912) algebraized topology.

— Lady of the Rings, and Some Lords —

Eduard Kummer (1810–1893) used algebra on Fermat's Last Theorem.

Richard Dedekind (1831–1916) discovered ideals.

David Hilbert (1862–1943): A geometry of tables, chairs, and beer mugs.

Emmy Noether (1882–1935) pulled it all together.

— Modern Algebra —

Solomon Lefschetz (1884–1972)
harpooned a whale.

Oscar Zariski (1899–1986)
refounded algebraic geometry.

Saunders Mac Lane (1909–2005)
attained a higher level of abstraction.

Alexander Grothendieck (1928–):
"As if summoned from the void."

Meanwhile he had taken up the serious study of mathematics, from about age 17. He quickly taught himself calculus. By his mid-20s he was publishing regularly in the *Cambridge Mathematical Journal*, with the encouragement of Duncan Gregory, the journal's first editor, whom I quoted earlier in this chapter. He began a correspondence with De Morgan in 1842, and De Morgan helped Boole to get a paper on differential equations published by the Royal Society.

When, in 1846,[104] the British government announced an expansion of higher education in Ireland, Boole's admirers—among them De Morgan, Cayley, and William Thomson (later Lord Kelvin, after whom the temperature scale is named)—agitated for Boole to be given a professorship at one of the new colleges. They were successful, and in 1849 Boole became professor of mathematics at Queen's College, Cork. He served in that position for 15 years, until a November day in 1864 when he walked the two miles from his house to the college in pouring rain, lectured in wet clothes, and caught a chill. His wife believed that a disease should be treated by methods resembling the cause, so she put George in bed and threw buckets of icy water over him. The result, as mathematicians say, followed.

I have been unable to find any source with an unkind word to say about George Boole. Even after discounting for the hyperbole of sympathetic biographers, he seems to have been a good man, to near the point of saintliness. He was happily married to a niece of Sir George Everest, the man the Himalayan mountain was named after. They had five daughters, the middle one of whom, Alicia Boole Stott, became a self-taught mathematician herself, did important work in multidimensional geometry, and lived to be 80, dying in World War II England.[105]

Boole's great achievement was the algebraization of logic—the elevation of logic into a branch of mathematics by the use of algebraic symbols. To illustrate Boole's method, here is an algebraized version of that syllogism I showed above.

Let us restrict our attention to the set consisting of all living things on Earth. This will be our "universe of discourse," though

Boole did not use this term, which was only coined in 1881, by John Venn. Denote this universe by 1. In the same spirit, use 0 to denote the empty set, the set having no members at all. Now consider all *mortal* living things, and denote this set by x. (Possibly $x = 1$; it makes no difference to the argument.) Similarly, use y to denote the set of all men and s the set whose only member is Socrates.

Two more notations: First, if p is a set of things and q is a set of things, I shall use the multiplication sign to show their intersection: $p \times q$ represents all things that are in both p and q. (There may be no such things. Then $p \times q = 0$.) Second, I shall use the subtraction sign to remove that intersection: $p - q$ represents all the things that are in p but not also in q.

Now I can algebraize my syllogism. The phrase "all men are mortal" can be restated as: "the set of living things that are a man AND not mortal is the empty set." Algebraically: $y \times (1 - x) = 0$. Multiplying out the parenthesis and applying *al-jabr*, this is equivalent to $y = y \times x$. (Translating back: "The set of all men is just the same as the set of all men who are mortal.")

"Socrates is a man" similarly algebraizes as $s \times (1 - y) = 0$, equivalent to $s = s \times y$. ("The set consisting just of Socrates is identical to the set whose members are at one and the same time Socrates and men." If Socrates were not a man, this would not be so; the latter set would be empty!)

Substituting $y = y \times x$ in the equation $s = s \times y$, I get $s = s \times (y \times x)$. By the ordinary rules of algebra, I can reposition the parentheses like this: $s = (s \times y) \times x$. But $s \times y$ I have already shown to be equal to s. Therefore $s = s \times x$, equivalent to $s \times (1 - x) = 0$. Translation: "The set of living things that are Socrates AND not mortal is empty." So Socrates is mortal.

In a much-quoted remark that first appeared in a 1901 paper, Bertrand Russell said: "Pure mathematics was discovered by Boole, in a work which he called *The Laws of Thought* (1854)." Russell goes on to let a little of the air out of that remark: "[I]f his book had really contained the laws of thought, it was curious that no one should ever have thought in such a way before. . . ." Russell was also speaking from a point of view he himself had arrived at about the relationship between mathematics and logic: that mathematics *is* logic, a belief no longer widely held today.

Most modern mathematicians would respond to Russell's remark by saying that what Boole had actually invented was not pure mathematics but a new branch of applied mathematics—the application of algebra to logic. The subsequent history of Boole's ideas bears this out. His algebra of sets was turned into a full logical calculus later in the 19th century by the succeeding generation of logicians: Hugh McColl, Charles Sanders Peirce (son of the Benjamin mentioned in §8.9), Giuseppe Peano, and Gottlob Frege. This logical calculus then flowed into the great stream of 20th-century inquiry known on math department lecture timetables as "Foundations," in which mathematical techniques are used to investigate the nature of mathematics itself.

Since that stream is not commonly considered to be a part of modern algebra, I shall not follow it any further. A history of algebra would not be complete without some account of Boolean algebra, though; so there he stands, George Boole of Lincoln, the man who married algebra to logic.

§10.6   Of the great generation of British mathematicians born in the first quarter of the 19th century, I have already given passing mention to Arthur Cayley in connection with the theory of matrices. That was by no means Cayley's only large contribution to algebra, though. He has a fair claim to having been the founder of modern abstract group theory, the topic of my next chapter. It is therefore convenient, as well as fair, to cover that aspect of Cayley's work here, before heading back to the European mainland.

The English word "group," in its modern algebraic meaning, first appears in two papers Cayley published in 1854, both under the same title: "On the Theory of Groups, as Depending on the Symbolic Equation $\theta^n = 1$." I am going to give a fuller account of early group theory in the next chapter. Here I only want to bring out a very useful feature of Cayley's 1854 presentation, as a sort of introduction to the topic.

Back in §9.2, when discussing determinants, I did some ad hoc work on the permutations of three objects and listed the six possibilities for such permutations. To explain Cayley's advance, I need to say much more about permutations and to introduce a good way of writing them. Three or four different ways to denote permutations have been in favor at one time or another, but modern algebraists seem to have definitely settled into the *cycle notation*, and that is the one I shall use from now on.

Cycle notation works like this. Consider three objects—apple, book, and comb—in three boxes labeled 1, 2, and 3. Consider this to be a "starting state": apple in box 1, book in box 2, comb in box 3. Define the "identity permutation" to be the one that changes nothing at all. If you apply the identity permutation to the starting state, the apple will stay in box 1, the book will stay in box 2, and the comb will stay in box 3.

Note—this is a point that often confuses beginning students— that a permutation acts on the contents of boxes, whatever those might be at any point. The permutation: "switch the contents of the first box with the contents of the second" is written in cycle notation as (12). This is read: "[The object in box] 1 goes to [box] 2; [the object in box] 2 goes to [box] 1." As the square brackets indicate, what a mathematician actually thinks when he sees that cycle notation is: "1 goes to 2, 2 goes to 1." Note the wraparound effect, the last number listed in the parentheses permuting to the first. That's why the notation is called cyclic!

Suppose we apply this permutation to the starting state. Then the apple will be in box 2, the book in box 1. If I *then* apply the do-nothing identity transformation, the apple remains in box 2, the book remains in box 1, and of course the comb remains in box 3. Using a multiplication sign to indicate the compounding of permutations, $(12) \times I = (12)$.

Suppose I did the (12) permutation from the starting state and then did it again. Doing it the first time, the apple goes to box 2 and the book to box 1. Doing it the second time, the book goes to box 2 and the apple to box 1. I am back at the starting state. In other words, $(12) \times (12) = I$.

Working like this, you can build a complete "multiplication table" of permutations on three objects. Consider the more complex permutation written as (132) in cycle notation. This is read: "[The object in box] 1 goes to [box] 3; [the object in box] 3 goes to [box] 2; [the object in box] 2 goes to [box] 1." Well, this permutation, applied to the starting state, would put the apple in box 3, the comb in box 2, and the book in box 1. ("1 goes to 3, 3 goes to 2, 2 goes to 1.") And if I *then* apply the (12) permutation, the comb would be in box 1, the book in box 2, and the apple in box 3—just as if, from the starting state, I had applied the permutation (13). To put it algebraically: $(132) \times (12) = (13)$.

As I said, you can build up an entire multiplication table this way. Here it is. To see the result of applying *first* a permutation from the list down the left-hand side, *then* one from the list along the top, just look along the appropriate row to the appropriate column.

	I	(123)	(132)	(23)	(13)	(12)
I	I	(123)	(132)	(23)	(13)	(12)
(123)	(123)	(132)	I	(13)	(12)	(23)
(132)	(132)	I	(123)	(12)	(23)	(13)
(23)	(23)	(12)	(13)	I	(132)	(123)
(13)	(13)	(23)	(12)	(123)	I	(132)
(12)	(12)	(13)	(23)	(132)	(123)	I

FIGURE 10-1 The Cayley table for the group S_3.

This is called a Cayley table. In fact, there is a table that closely resembles this one on page 6 of the first of those 1854 papers. Note that the compounding of permutations is noncommutative, as can be seen from the fact that this table is *not* symmetrical about the lead diagonal (top left to bottom right).

These six permutations of three objects, together with the rule for combining the permutations as defined by that table, are an example of a group. This particular group is important enough to have its own symbol: S_3.

S_3 is not the only group with six elements. There is another one. Consider the sixth roots of unity. Using ω, as usual, to denote the first cube root of unity, the sixth roots are: $1, -\omega^2, \omega, -1, \omega^2$, and $-\omega$. (See §RU.5 for a reminder.) An ordinary multiplication table for these six numbers looks like this:

	1	$-\omega^2$	ω	-1	ω^2	$-\omega$
1	1	$-\omega^2$	ω	-1	ω^2	$-\omega$
$-\omega^2$	$-\omega^2$	ω	-1	ω^2	$-\omega$	1
ω	ω	-1	ω^2	$-\omega$	1	$-\omega^2$
-1	-1	ω^2	$-\omega$	1	$-\omega^2$	ω
ω^2	ω^2	$-\omega$	1	$-\omega^2$	ω	-1
$-\omega$	$-\omega$	1	$-\omega^2$	ω	-1	ω^2

FIGURE 10-2 The Cayley table for the group C_6.

That one *is* commutative, as you would expect, since I am just multiplying ordinary (I mean, ordinary complex) numbers. Its name is C_6, the cyclic group with six elements.

Those are examples of the two groups that have six elements— groups of order 6, to use the proper term of art. What makes them groups? Well, certain features of the multiplication tables are critical.

Note, for instance, that the unity (I in the case of the first group, 1 in the second) appears precisely once in each row and once in each column of the multiplication table.

The most important word in that last paragraph is the third one: "examples." In the minds of mathematicians, S_3 and C_6, the two groups of order 6, are perfectly abstract objects. If you were to replace the symbols for the six permutations of three objects with 1, $\alpha, \beta, \gamma, \delta$, and ε, and go through the Cayley table replacing each symbol with its appropriate Greek letter, that table would stand as a definition of the abstract group, with no reference to permutations at all. That is, in fact, precisely how Cayley does it in his 1854 papers. The group of permutations on three objects is an *instance* of the abstract group S_3, just as the justices of the United States Supreme Court are an instance of the abstract number 9. Similarly for the sixth roots of unity. With the operation of ordinary multiplication, they form an instance of C_6. You could replace them by 1, α, β, etc., make appropriate replacements in the second table above, and there is a perfectly abstract definition of C_6, without any reference to roots of unity.

That was Cayley's great achievement, to present the idea of a group in this purely abstract way. For all Cayley's insight, though, and fully acknowledging the great conceptual leap these 1854 papers represent, Cayley could not detach his subject completely from its origins in the study of equations and their roots. In a sort of backward glance to those origins, he appended this footnote on the second page of the first paper:

> The idea of a group as applied to permutations or substitutions is
> due to Galois, and the introduction of it may be considered as mark-
> ing an epoch in the progress of the theory of algebraic equations.

Cayley was quite right. It is now time to go back a little, to take another pass at the middle quarters of the 19th century, and to meet algebra's only real romantic hero, Évariste Galois.

Part 3

LEVELS OF ABSTRACTION

Math Primer

FIELD THEORY

§FT.1 "FIELD" AND "GROUP" ARE the names of two mathematical objects that were discovered in a series of steps during the early 19th century.

A field is a more complicated thing than a group, so far as its internal structure goes. For this reason, textbooks of algebra usually introduce groups first and then advance to fields, even though there is a sense in which a field is a more commonplace kind of thing than a group and therefore easier to comprehend. Being simpler, groups also have a wider range of applicability, so that on the whole, group theory is more challenging to the pure algebraist than field theory.[106]

For these reasons, and also to make Galois' discoveries more accessible, I am going to describe fields here in this primer, before going into more detail about groups in the chapter that follows.

§FT.2 A field is a system of numbers (or other things—but numbers will serve for the time being) that you can add, subtract, multiply, and divide to your heart's content. No matter how many of the four basic operations you do, the answer will always be some other

number in the same field. That is why I said that a field is a commonplace sort of thing. When you are in a field, you are doing basic arithmetic: +, −, ×, ÷. If you want a visual mnemonic for the algebraic
concept of a field, just imagine the simplest kind of pocket calculator,
with its four operation keys: +, −, ×, ÷.

There are certain rules to be followed, none of them very surprising. I have already mentioned the closure rule: Results of arithmetic
operations stay within the field. You need a "zero" that leaves other
numbers unchanged when you add it and a "one" that leaves them
unchanged when you multiply by it. Basic algebraic rules must apply:
$a \times (b + c)$ always equal to $a \times b + a \times c$, for instance. Both addition and
multiplication must be commutative; we have no truck with non-
commutativity in fields. Hamilton's quaternions are therefore not a
field, only a "division algebra."

Neither $\mathbb{N}$ nor $\mathbb{Z}$ is a field, since dividing two whole numbers
may not give a whole-number answer. The family of rational numbers $\mathbb{Q}$ does form a field, though. You can add, subtract, multiply, and
divide as much as you like without ever leaving $\mathbb{Q}$. It's a field. There is
a sense in which $\mathbb{Q}$ is the most important, the most *basic*, field. The
real numbers $\mathbb{R}$ form a field, too. So do the complex numbers $\mathbb{C}$,
using the rules for addition, subtraction, multiplication, and division
that I gave at the beginning of this book. We therefore have three
examples of fields already to fix our ideas on: $\mathbb{Q}$, $\mathbb{R}$, and $\mathbb{C}$.

Are there any other fields besides $\mathbb{Q}$, $\mathbb{R}$, and $\mathbb{C}$? There certainly
are. I am going to describe two common types. Then I shall put the
two types together to lead into Galois theory and group theory. Then,
as a footnote, I shall mention a third important type of field.

§FT.3   The first other type of field—other than the familiar $\mathbb{Q}$, $\mathbb{R}$,
and $\mathbb{C}$—is the *finite* field. $\mathbb{Q}$, $\mathbb{R}$, and $\mathbb{C}$ all have infinitely large mem-
berships. There is an infinity of rational numbers; there is an infinity
of real numbers; and there is an infinity of complex numbers.

Here is a field with only three numbers in it, which for convenience I shall call 0, 1, and 2, though if you find this leads to too much
confusion with the more usual integers with those names, feel free to
scratch out my 0, 1, 2, and replace them with any other symbols you
please: perhaps "Z" for the zero, "I" for the one, and "T" for the third
field element. It will *not* be the case in this field, for example, that
$2 + 2 = 4$. In this field, $2 + 2 = 1$. Here, in fact, are the complete addition and multiplication tables for my sample finite field, whose
name is Γ_3.

+	0	1	2		×	0	1	2
0	0	1	2		0	0	0	0
1	1	2	0		1	0	1	2
2	2	0	1		2	0	2	1

FIGURE FT-1 The field F_3.

Note some points about this field. First, since the additive inverse
("negative") of 1 is 2, and vice versa, there is not much point in talking about subtraction. ". . . −1" can always be replaced by ". . . +2," and
vice versa.[107] Same with division. Since the multiplicative inverse ("reciprocal") of 2 is 2 (because $2 \times 2 = 1$), a division by 2 can always be
replaced by a multiplication by 2, with exactly the same result! Division by 1 is trivial, and division by zero is never allowed in fields.

Is there a finite field for every natural number greater than 1? No.
There are finite fields only for prime numbers and their powers. There
are finite fields with 2, 4, 8, 16, 32, . . . members; there are finite fields
with 3, 9, 27, 81, 243, . . . members; and so on. There is, however, no
finite field with 6 members or 15 members.

Finite fields are often called Galois fields, in honor of the French mathematician Évariste Galois, whom we shall meet presently in the main text.

§FT.4   The second other type of field is the *extension field*. What we do here is take some familiar field—very often $\mathbb{Q}$—and append one extra element to it. The extra element should, of course, be taken from outside the field.

Suppose, for example, we append the element $\sqrt{2}$ to $\mathbb{Q}$. Since $\sqrt{2}$ is not in $\mathbb{Q}$, this should be just the kind of thing I am talking about. If I now add, subtract, multiply, and divide in this enlarged family of numbers, I get all numbers of the form $a+b\sqrt{2}$, where a and b are rational numbers. The sum, difference, product, and quotient of any two numbers of this kind are other numbers of the same kind. The rules for addition, subtraction, multiplication, and division in fact look rather like the rules for complex numbers. Here, for example, is the division rule:

$$\left(a+b\sqrt{2}\right)\div\left(c+d\sqrt{2}\right) \;=\; \frac{ac-2bd}{c^2-2d^2} + \frac{bc-ad}{c^2-2d^2}\sqrt{2}$$

This is a field. I have extended the field of rational numbers $\mathbb{Q}$ by just appending the one irrational number $\sqrt{2}$. This gives me a new field.

Note that this new field is not $\mathbb{R}$, the field of real numbers. All kinds of real numbers are *not* in it: $\sqrt{3}$, $\sqrt[5]{12}$, π, and an infinite host of others. The only numbers that *are* in it are (i) all the rational numbers, (ii) $\sqrt{2}$, and (iii) any number I can get by combining $\sqrt{2}$ with rational numbers via the four basic arithmetic operations.

Why would I want to go to all this trouble to extend $\mathbb{Q}$ by such a teeny amount? To solve equations, that's why. The equation $x^2-2=0$

has no solutions in $\mathbb{Q}$, as Pythagoras discovered to his alarm and distress. In this new, slightly enlarged field, it *does* have solutions, though: $x = \sqrt{2}$ and $x = -\sqrt{2}$. By extending fields judiciously, I can solve equations I couldn't solve before.

Notice an interesting and important thing: The extended field is a vector space over the original field $\mathbb{Q}$. An example of two linearly independent vectors would be the numbers 1 and $\sqrt{2}$. These would, in fact, make an excellent basis (see §VS.3) for the vector space. Every other vector—every number of the form $a + b\sqrt{2}$, with a and b both rational—can be expressed in terms of them. Considered in this way, as a vector space, the extended field is two-dimensional.

The field you get by appending an irrational number to $\mathbb{Q}$ will not always be two-dimensional. If, for example, you were to append $\sqrt[3]{2}$, the extension field would be three-dimensional, with the three vectors 1, $\sqrt[3]{2}$, $\sqrt[3]{4}$ as a suitable basis. Here, just to show how quickly things can get out of control, is the rule for division in this field:

$$\left(a + b\sqrt[3]{2} + c\sqrt[3]{4}\right) \div \left(f + g\sqrt[3]{2} + h\sqrt[3]{4}\right)$$

$$= \frac{af^2 - 2agh + 2bg^2 - 2bfh + 4ch^2 - 2cfg}{f^3 + 2g^3 + 4h^3 - 6fgh}$$

$$+ \frac{2ah^2 - afg + bf^2 - 2bgh + 2cg^2 - 2cfh}{f^3 + 2g^3 + 4h^3 - 6fgh}\sqrt[3]{2}$$

$$+ \frac{ag^2 - afh + 2bh^2 - bfg + cf^2 - 2cgh}{f^3 + 2g^3 + 4h^3 - 6fgh}\sqrt[3]{4}$$

§FT.5 I am now going to put the previous two sections together and solve some quadratic equations in my 0, 1, 2 field. The advantage of finite fields, you see, is that you can write down *all possible* quadratic equations!

First things first. Here are all possible *linear* equations in my 0, 1, 2 field, with their solutions, which you can check if you like against the addition and multiplication tables for that field.

Equation	Solution
$x = 0$	$x = 0$
$2x = 0$	$x = 0$
$x + 1 = 0$	$x = 2$
$x + 2 = 0$	$x = 1$
$2x + 1 = 0$	$x = 1$
$2x + 2 = 0$	$x = 2$

In fact, I have even made too much of that. The first two equations are not really interesting. Of course the solution of $2x = 0$ is $x = 0$! A bit less obvious, neither are the last two very interesting. Their left-hand sides factorize to, respectively, $2(x + 2)$ and $2(x + 1)$, so they are really just the third and fourth equations over again in light disguise. (Remember that in this field $2 \times 2 = 1$.) Only the middle two equations are really of any interest.

On to quadratic equations. This time I shall discard the uninteresting ones in advance. Here are all the interesting quadratic equations with coefficients in F_3. For extra points, I have factorized them, too.

Equation	Factorizes as	Solutions
$x^2 + 1 = 0$	won't factorize	no solutions
$x^2 + 2 = 0$	$(x + 1)(x + 2)$	$x = 1, x = 2$
$x^2 + x + 1 = 0$	$(x + 2)^2$	$x = 1$
$x^2 + x + 2 = 0$	won't factorize	no solutions
$x^2 + 2x + 1 = 0$	$(x + 1)^2$	$x = 2$
$x^2 + 2x + 2 = 0$	won't factorize	no solutions

An equation that has no solutions in the field I am working with is called *irreducible*. (Compare Endnote 34.) You can see that of the six interesting equations with coefficients in the 0, 1, 2 field, three are irreducible.

See what I have done? I have re-created in miniature the situation you get with "normal" quadratic equations—except that, instead of an infinity of equations to worry about, in this field there are only six: three with solutions, three irreducible. In normal arithmetic the equation $x^2 - 2 = 0$ has no solutions in $\mathbb{Q}$ because $\sqrt{2}$ is not a rational number. Similarly, the equation $x^2 + 1 = 0$ has no solutions in $\mathbb{Q}$, or even in $\mathbb{R}$, because $\sqrt{-1}$ is not in either $\mathbb{Q}$ or $\mathbb{R}$.

§FT.6   Can we extend the 0, 1, 2 field so that those irreducible equations have solutions? Yes, we can. Let's invent a new number—I shall just call it *a*—that satisfies that first equation: $a^2 + 1 = 0$. Adding 2 to each side, $a^2 = 2$. (So you could call *a* a square root of 2. Since this 2 isn't really behaving altogether like a regular 2, though, I won't write *a* as $\sqrt{2}$. I'll just leave it incognito as *a*.) And now all the equations can be solved:

Equation	Factorizes as	Solutions
$x^2 + 1 = 0$	$(x + 2a)(x + a)$	$x = a, x = 2a$
$x^2 + 2 = 0$	$(x + 1)(x + 2)$	$x = 1, x = 2$
$x^2 + x + 1 = 0$	$(x + 2)^2$	$x = 1$
$x^2 + x + 2 = 0$	$(x + 2a + 2)(x + a + 2)$	$x = a + 1, x = 2a + 1$
$x^2 + 2x + 1 = 0$	$(x + 1)^2$	$x = 2$
$x^2 + 2x + 2 = 0$	$(x + 2a + 1)(x + a + 1)$	$x = a + 2, x = 2a + 2$

We just needed to add that one element *a* to the field, and we can solve all quadratic equations. And all addition, subtraction, multipli-

cation, and division in the extended field, which is commonly denoted by $F_3(a)$, involve nothing more than linear expressions in a. If a
multiplication results in a^2, you can at once replace it by 2 because
$a^2 = 2$. Here is the multiplication table for the extended field. (The
addition table is less exciting, though you should feel free to construct it if you want to.)

×	0	1	2	a	$2a$	$1+a$	$1+2a$	$2+a$	$2+2a$
0	0	0	0	0	0	0	0	0	0
1	0	1	2	a	$2a$	$1+a$	$1+2a$	$2+a$	$2+2a$
2	0	2	1	$2a$	a	$2+2a$	$2+a$	$1+2a$	$1+a$
a	0	a	$2a$	2	1	$2+a$	$1+a$	$2+2a$	$1+2a$
$2a$	0	$2a$	a	1	2	$1+2a$	$2+2a$	$1+a$	$2+a$
$1+a$	0	$1+a$	$2+2a$	$2+a$	$1+2a$	$2a$	2	1	a
$1+2a$	0	$1+2a$	$2+a$	$1+a$	$2+2a$	2	a	$2a$	1
$2+a$	0	$2+a$	$2+2a$	$2+2a$	$1+a$	1	$2a$	a	2
$2+2a$	0	$2+2a$	$1+a$	$1+2a$	$2+a$	a	1	2	$2a$

FIGURE FT-2 The multiplication table for $F_3(a)$.

§FT.7   There we have some highly concentrated essence of Galois
theory. We have an equation whose coefficients belong to a certain
field but whose solutions can't be found in that field. In order to encompass those solutions, we extend our coefficient field to a larger
field—call it the solution field. The issue Galois was concerned with,
the issue of what form the solutions of our equation will take, *de-*
pends on the relationship between these two fields, the coefficient field
and the solution field.

That was Galois' great insight. His discovery was that this relationship can be expressed in the language of group theory, which, in 1830, meant the language of permutations.

Galois found that, for any given equation, we need to consider certain *permutations* of the solution field. The solution field, like my $F_3(a)$ above, is in general bigger than the coefficient field (F_3 in my example). Now, among all useful permutations of the solution field, there is a subfamily of permutations *that leave the coefficient field unchanged*. That subfamily forms a group, which we call the Galois group of the equation. All questions about the solvability of the equation translate into questions about the structure of that group.

In the case of the equation I began this section with, the equation $x^2 + 1 = 0$, with coefficients understood to be taken from the minifield F_3, the Galois group is a rather simple one, with only two members. One of those members is the identity permutation I, which leaves everything alone. The other is the permutation that exchanges the two solutions, sending a to $2a$ and $2a$ to a. This permutation—let's call it P—acts on the whole of $F_3(a)$, of course. Using an arrow to indicate "permutes to," it acts like this: $0 \rightarrow 0, 1 \rightarrow 1, 2 \rightarrow 2, a \rightarrow 2a, 2a \rightarrow a, 1 + a \rightarrow 1 + 2a, 1 + 2a \rightarrow 1 + a, 2 + a \rightarrow 2 + 2a, 2 + 2a \rightarrow 2 + a$.

Here is a "multiplication" table for the Galois group of the equation $x^2 + 1 = 0$ over the coefficient field F_3. Multiplication here means the compounding of permutations—doing one permutation, then doing the other.

	I	P
I	I	P
P	P	I

FIGURE FT-3 The multiplication table for the Galois group of $x^2 + 1 = 0$.

§FT.8   That is a grossly oversimplified account of Galois theory, of course.[108] It is all very well to speak of permuting things like F_3 or $F_3(a)$, which have only three and nine members, respectively. What had been vexing algebraists for all those centuries was the solution of polynomials with coefficients in $\mathbb{Q}$, a field with *infinitely many* members. How do you permute *that*?

I hope to make this a little clearer as I proceed. I doubt I can make things *much* clearer, though. Galois theory is a difficult and subtle branch of higher algebra, not easily accessible to the non-mathematician. If you can keep in your mind the fact that a polynomial with coefficients in a certain field may have roots in a bigger field, that the relationship between these two fields, the bigger one and the smaller one, can be expressed in the language of group theory, and that every question about solving a polynomial equation can thereby be translated into a question about group theory, you will have grasped the essence of Galois's achievement.

§FT.9   Before leaving the topic of fields, I had better add one more type of field, mainly by way of apology. In discussing the work of the 18th-century algebraists, I used the word "polynomial" a bit indiscriminately, just for the sake of simplicity. Some of those usages should really have been not "polynomial" but "rational function."

A rational function is the ratio of two polynomials, like this:

$$\frac{2x^2 + x - 3}{3x^3 + 2x^2 - 4x - 1}$$

Since, with a little labor, any two such functions can be added, subtracted, multiplied, or divided, they form a field.

Note that a field of rational functions "depends on" another field, the field from which the coefficients of the polynomials are taken. A field of rational functions can in fact be viewed from the perspective of field extensions, as described above. I start with my coefficient field, whatever it is. Then I append the symbol x and permit all possible additions, subtractions, multiplications, and divisions. This generates the rational-function field. The only difference between this and my previous examples of field extensions is that in those cases I had a better handle on the thing I was appending. I knew that its square, or its cube, was 2. This allowed me to do a lot of simplification on the field arithmetic. Here I don't know *anything* about *x*. It's just a symbol—an unknown quantity, if you like. . . .

Chapter 11

PISTOLS AT DAWN

§11.1 MATHEMATICS IS, LET'S FACE IT, a dry subject, with little in the way of glamour or romance. The story of Évariste Galois has therefore been made much of by historians of math.

A little too much, perhaps. The facts of the Galois story, to the degree that they are known, have been surrounded by a fog of myth, error, speculation, and agenda-peddling. Best known in English is the chapter "Genius and Stupidity" in E. T. Bell's classic *Men of Mathematics*, which tells the story of a legion of fools persecuting an ardently idealistic genius who spends his last desperate night on earth committing the foundations of modern group theory to paper. Bell's story is certainly false in details and probably in the character it draws for Galois. Tom Petsinis's 1997 novel *The French Mathematician*, though written in a style not at all to my taste and with an ending that strikes me as improbable, is sounder on the facts of Galois' case. Best of all for an hour or so of reading is the Web site by cosmologist and amateur math-historian Tony Rothman, which weighs all the sources very judiciously.[109]

Galois died in a duel at the age of 20 years and 7 months. The duel, fought with pistols, naturally took place at dawn. Galois, apparently sure that he was going to be killed—perhaps even wishing for it—sat up the night before, writing letters. Some were to political friends, antimonarchist republicans like himself. One, with some annotations to his mathematical work, was to his friend Auguste Chevalier.

The reason for the duel is unclear. It was either political or amorous, possibly both. "I have been provoked by two patriots . . . ," Galois wrote in one of those letters. In another, however, he says: "I die the victim of an infamous coquette. . . ." Galois had been involved in the extreme-republican politics that flourished in Paris around the time of the 1830 uprising—the uprising portrayed in Victor Hugo's *Les Misérables*. He was also suffering from unrequited love.

Romantic enough, to be sure. As usually happens in these cases, a close look at the circumstances replaces some of the romance with pathos and squalor. Galois's story is certainly a sad one, though, and the fact that his own awkward personality seems to have played a large part in his misfortunes does little to diminish the sadness.

§11.2   The France of 1830 was not a happy nation. The king, Charles X, of the Bourbon dynasty restored by the allies after the defeat of Napoleon, was old and reactionary. Down at the other end of society, rapid urbanization and industrialization were turning much of Paris into a horrible slum, in which hundreds of thousands dwelt in misery and near-starvation. This was the Paris drawn by Balzac and Victor Hugo, where a thrusting, materialistic bourgeoisie dwelt alongside a seething underclass. The employment prospects of the latter were at the mercy of untamed business cycles; their miseries were alleviated only by occasional charity.

In 1830, there was a recession. Bread prices soared, and more than 60,000 Parisians had no work. In July, barricades went up; the mob took control of the city; Charles X was forced to flee the country. Louis Philippe, the Duke of Orléans, from a distant branch of the Bourbons, was chosen by progressive-bourgeois parliamentary deputies to be the new king—"the July monarch." Amiable and unpretentious, Louis Philippe was something of a limousine liberal. A radical element was coming up in French politics, however, and they could not be satisfied by any mere liberal. The 1830s were punctuated by insurrections, including a major one in Paris in 1831. These were tense years, when hot-headed young men with strong opinions could reasonably expect to find themselves watched by the police, perhaps to do some brief prison time.

§11.3   Évariste Galois was born in October 1811 in the little town of Bourg-la-Reine, just south of Paris—it is now a suburb—on the road to Orléans. Galois *père* was a liberal, an anticlericalist and antiroyalist. He had been elected mayor of the town in 1815, during Napoleon's last days as emperor, the "hundred days" that ended at Waterloo. After the monarchy was restored, this elder Galois took an oath of loyalty to the Bourbons—not from any change of heart but to prevent a real royalist from getting his job.

The first comments we have about Galois' character come from his Paris schoolmasters. They show a youth who was clever but introverted, not well organized in his work, and not willing to listen to advice. Tony Rothman notes that: "The words 'singular,' 'bizarre,' 'original' and 'withdrawn' would appear more and more frequently during the course of Galois' career at Louis-le-Grand. His own family began to think him strange." Rothman adds, however, that the opinions of Galois' teachers were far from unanimous and that his schooldays were by no means the nightmare of uncomprehending persecution described by E. T. Bell.

In July of 1829—Galois was not yet 18—his father, who had been enduring a campaign of slander by a malicious local priest, committed suicide in a Paris apartment just a few yards from Évariste's school. The event caused Galois intense and lasting distress. Just a few days later he had to attend a *viva voce* examination for entry to the very prestigious École Polytechnique, whose faculty included Lagrange, Laplace, Fourier, and Cauchy. Galois failed the exam through tactlessness, possibly rising to the level of willful arrogance. At one point he responded to a request to prove some mathematical statement by saying that the statement ought to be perfectly obvious. A few months later—we are now in early 1830, Galois was 18½—he was accepted into the less prestigious École Preparatoire, essentially a teacher training college (now called the École Normale).

The first version of Galois' theory on the solution of equations—the paper that E. T. Bell implies was scribbled out frantically the night before Galois died—had actually been submitted to the Academy of Sciences a few weeks before Galois *père* killed himself. Cauchy was appointed referee for the paper. Bell (and, to be fair, everyone else, until recent researches in the Academy's archives turned up exculpatory evidence) believed that Cauchy just lost or ignored the paper. To the contrary, the great man seems to have thought highly of it. It is likely he suggested that Galois polish it a little and submit it for the Academy's grand prize in mathematics. At any rate, whether at Cauchy's suggestion or not, Galois did just that, submitting the paper a second time in February 1830 to Joseph Fourier, secretary of the Academy. Alas, Fourier died on May 16.

Cauchy might yet have rescued Galois from obscurity. This was the year of revolution, though, and the new liberal regime of Louis Philippe was hard for Cauchy to stomach. He was in any case a man of strong principles. Having sworn an oath of loyalty to Charles X, he did not feel he could now swear one to Louis Philippe. He might have just resigned his chairs and retired to some private provincial position (he was 40 at this point). Instead, he exiled himself, staying out of France for eight years. There is no good explanation for Cauchy's self-imposed exile, other than the penchant for "quixotic behavior" noted by Freudenthal (§7.5).

Galois himself did not take part in the July revolution. Knowing that his student body contained a large radical element, the director of the École Preparatoire locked the students in, so that they could not take part in the street fighting. Galois was, though, sufficiently free with his radical opinions to get himself expelled from the college.

That was at the beginning of January 1831. The last 17 months of Galois' life then proceeded as follows:

> January 4, 1831—Expelled from the École Preparatoire. Galois seems to have spent the next four months trying to make a living by teaching mathematics privately in Paris, while hanging out with other young people of extremist republican sympathies.
>
> January 17—Submitted a third version of the memoir on the solution of equations to the Academy. Siméon-Denis Poisson was to be the referee.
>
> May 9—Attended a rowdy republican banquet at which he seemed, when proposing a toast, to have threatened the life of Louis-Philippe. Galois was arrested the next day.
>
> June 15—Tried but acquitted, probably on account of his youth.
>
> July 14 (Bastille Day)—Got arrested again, with his friend Ernest(?) Duchâtelet, for wearing the uniform of the banned Artillery Guard. Also, apparently, for being armed. Galois is reported to have been in possession of a loaded rifle, "several pistols," and a dagger.
>
> (Galois was imprisoned from July 14, 1831, to April 29, 1832. The conditions of imprisonment do not seem to have been very arduous, though. The prisoners were, for example, frequently drunk.)
>
> October—Received a letter of rejection from Poisson at the Academy. He had found Galois' paper too difficult to follow, though he was not condemnatory, and suggested an improved presentation.
>
> March 16, 1832—Transferred with other inmates from the prison to a sanatorium, to protect them from the cholera epidemic then sweeping Paris. The sanatorium served as an "open prison," and Galois had considerable freedom to come and go. Here he fell in love with Stéphanie Dumotel, daughter of one of the resident physicians. His love, however, was not requited.
>
> April 29—Galois freed.
>
> May 14—The date on a rejection letter Galois seems to have received from Stéphanie.

May 25—The date on a letter written by Galois to his friend Auguste
  Chevalier, telling of a broken love affair.
May 30—The fatal duel.

The precise circumstances of the duel, and indeed of Galois' last
few days of life, are mysterious and will probably always remain so. It
is likely that Galois had, in some sense, given up on life. The death of
his father, the rejection of his paper (following on the misfortunes of
its previous submission), his unrequited love for Stéphanie, his own
small-to-nonexistent employment prospects, the months of confine-
ment, the wretchedness of Paris during the epidemic—it was all too
much for him.

On June 4, a Lyons newspaper printed a brief report on the duel
that makes it appear to be a competition between two old friends for
some woman's favors, settled by a sort of Russian roulette. "The pistol
was the chosen weapon of the adversaries, but because of their old
friendship they could not bear to look at one another. . . ." The newspaper identified Galois' adversary only as "L. D." Presumably the
woman was Stéphanie, but who is L. D.? The D could, under the pre-
vailing standards of orthography, stand for either Duchâtelet or
Perscheux d'Herbinville, another republican acquaintance of Galois.
Neither is known to have had a forename beginning with L, but then
neither is known not to have. French parents can be generous with
forenames.

Galois' brother and friends copied out his papers and circulated
them to big-name mathematicians of the day, including Gauss, but
with no immediate result. At length, 10 years after the fatal duel, the
French mathematician Joseph Liouville took an interest in the papers. He announced Galois' main result to the French Academy in
1843 and published all of Galois' papers three years later in a math
journal he had founded himself.[110] Only then did the name of Évariste
Galois become known to the larger mathematical community.

§11.4   What was the nature of Galois' work on the solution of equations that made it so important to the development of algebra? I shall give a brief account here, but I shall use modern language, not the language Galois himself used.

In Figure 10-1, I displayed a Cayley table—that is, a "multiplication" table—for the permutations of three objects. There are six possible permutations, and they can be compounded (first do this one, then do that one) according to the table. After that, in Figure 10-2, I showed another table, the one for multiplication of the sixth roots of unity. I said that these two tables illustrated the two groups having six members—groups of order 6.

Those tables show the essential features of abstract group theory. A group is a collection of objects—permutations, numbers, *anything*—together with a rule for combining them. The rule is most often represented as multiplication, though this is just a notational convenience. If the objects are not numbers, it can't be *real* multiplication.

To qualify as a group, the assemblage (objects plus rule) must obey the following principles or axioms:

Closure. The result of compounding two of the elements must be another one of them—must "stay within the group."

Associativity. If a, b, and c are any elements of the group and $\times$ is the rule for compounding them, then $a \times (b \times c) = (a \times b) \times c$ always. With this rule we can unambiguously compound three or more elements of the group.

Existence of a unity. There is some element of the group that, when compounded with any other element, leaves the other element unchanged. If we call this special element 1, then for every element a in the group it is the case that $1 \times a = a$.

Every element has an inverse. If a is any element of the group, I can find an element b for which $b \times a = 1$. This element is called the inverse of a and is frequently written as a^{-1}.

This highly abstract way of defining a group by means of axioms, using the language of set theory—"elements," "compounding"—is typical of the 20th-century *axiomatic* approach I have mentioned. This approach was of course not available to Galois, who expressed his ideas in terms of the particular properties of permutation groups.

The number of elements in a group is called its *order*. You can easily check that both those six-member groups in §10.6 satisfy the axioms for a group. What is not so easy to check is that there are no other groups of order 6. I mean *abstract* groups, of course—there are many other instances of six things that behave groupily (I shall produce one in a moment), but their rules of combination all follow the pattern in one or another of the Cayley tables of §10.6. Those multiplication tables offer the only two possible patterns for groups of six elements. There are no others. That is, there are only two abstract groups of order 6, though each one has numerous illustrating instances. Cayley, in his 1854 papers, listed all the groups of order up to 6. Nowadays we of course know far more groups. Figure 11-1 shows the number of abstract groups of order n, for n from 1 to 15.

How can we find the number of groups for any given order n? There is no general method and no formula. There are, however, some things to be noticed. If n is a prime number, for example, there seems never to be more than one group of order n. That is correct. For any prime number p, the only group of order p is C_p, the group illustrated by the pth roots of unity with ordinary multiplication and known technically as the cyclic group of order p.

n	1	2	3	4	5	6	7	8	9	10	11	12	13	14	15
Number of groups	1	1	1	2	1	2	1	5	2	2	1	5	1	2	1

FIGURE 11-1 The number of groups of order n, for n from 1 to 15.

Here are the two groups of order 4. Both were found by Cayley. I am just going to use arbitrary symbols α, β, and γ for their elements (other than the unity, which I shall write as 1).

×	1	α	β	γ
1	1	α	β	γ
α	α	β	γ	1
β	β	γ	1	α
γ	γ	1	α	β

×	1	α	β	γ
1	1	α	β	γ
α	α	1	γ	β
β	β	γ	1	α
γ	γ	β	α	1

FIGURE 11-2 The two abstract groups of order 4, C_4 and $C_2 \times C_2$.

These two groups both have names. The one on the left is named C_4. It is the cyclic group of order 4, illustrated by the fourth roots of unity, as you can see by setting α, β, γ equal to i, -1, and $-i$, respectively. The one on the right is named $C_2 \times C_2$, or the Klein 4-group. Both are commutative.[111]

Looking at that left-hand group C_4, if I tell you that the groups of order 3, 5, and 7 are named C_3, C_5, and C_7, respectively, you should be able to write out their multiplication tables very easily. The only group of order 2 is the one in my Figure FT-3. The order-1 group is of course trivial, included only for completeness. So now you know all the groups of order up to 7, which puts you slightly ahead of Arthur Cayley.

§11.5   Galois' great insight concerned the *structure* of abstract groups. Look at that group on the right in Figure 11-2, the Klein 4-group. The two elements 1 and α form a little group within the group—a *subgroup*, of order 2. So, for that matter, do 1 and β or 1 and γ. If you now look at the left-hand group in Figure 11-2, though, the one I introduced as C_4, you will see that 1 and β form a little group within a group, but 1 and α or 1 and γ do not. This is what I mean by structure. These groups-within-groups, these *subgroups*, play a key role in group theory.

Look back to that multiplication table for the group S_3, the group of permutations on three objects, in Figure 10-1. A number of subgroups are present. There is a subgroup of order 3, consisting of I, (123), and (132) (all the *even* permutations, please note). Then there are three subgroups of order 2: the one consisting of I and (23), the one consisting of I and (13), and the one consisting of I and (12). Each of these four subgroups forms its own happy little self-contained unit. Within it you can multiply as much as you like, take inverses as often as you like, and you will never be dragged outside the subgroup. In the second order-6 subgroup (Figure 10-2), the one named C_6, there is one subgroup of order 3, consisting of the cube roots of unity (1, ω, and ω^2), and there is one subgroup of order 2, consisting of the square roots of unity (1 and -1).

The first great theorem about group structure was Lagrange's theorem, which I mentioned in §7.4: The order of a subgroup divides the order of a group exactly. The quotient of this division is called the *index* of the subgroup. Lagrange's theorem forbids a fractional index. We may find subgroups of order 2 or 3 (index 3 or 2, respectively) in a group of order 6, but we can be sure we shall never find a subgroup of order 4 or 5 because neither 4 nor 5 divides into 6.

Galois added a key concept to the notion of group structure, the concept of what we nowadays call a *normal subgroup*. Here is a very brief account, using for illustration the group S_3, and the subgroup made up of I and (12), which I'll call H, and which is, of course, an instance of the one and only abstract group of order 2, shown in Figure FT-3.

Pick an element of the main group—one either inside or outside the subgroup; it doesn't matter. I'll use (123). Multiply I and (12) in turn by this element. Do the multiplication "from the left," applying first (123) and then the other permutation, as in §10.6. Result: a set—not a subgroup, just a set—consisting of (123) and (23). This is called a "left coset" of *H*. Repeat what you just did with (123), using every other element of the main group S_3. This gives you an entire family of left cosets. There is the one I just showed, {(123), (23)}. (Note: Those curly brackets are the usual way to denote a set. The set consisting of London, Paris, and Rome is written mathematically as {London, Paris, Rome}.) Then there is this one: {(132), (13)}. And there is this one: {*I*, (12)}, which is just *H* itself. (Since S_3 has order 6, you might have expected six members in the left-coset family, but they turn out to be equal in pairs.)

Repeat the entire process, but this time multiply "from the right." This gives you a family of "right cosets": {(123), (13)}, {(132), (23)}, and {*I*, (12)}. Now, if the family of left cosets is identical to the family of right cosets, then *H* is a normal subgroup. In my example the two families are not identical, so this particular *H* is not a normal subgroup of S_3. It's just a plain-vanilla subgroup. However, the order-3 subgroup of S_3, the one consisting of all the even permutations, *is* a normal subgroup. I leave you to check that. And notice one thing that follows from this definition: If a group is commutative, then *every* subgroup is normal.

Galois showed that to any polynomial equation of degree *n* in one unknown

$$x^n + px^{n-1} + qx^{n-2} + \ldots = 0,$$

we can, by studying the relationship between the field of coefficients and the field of solutions (see §FT.8), associate a group. If this "Galois group" of the equation has a structure that satisfies certain conditions, in which the concept of a normal subgroup is of central importance, then we can express the equation's solutions using only addition, subtraction, multiplication, division, and extraction of roots. If it doesn't, we can't. If *n* is less than 5, the equation's Galois group will *always* have the appropriate structure. If *n* is 5 or more, it may or may not have, depending on the actual numerical values of *p*, *q*, and the other coefficients. Galois uncovered those group-structural conditions, thereby supplying a definitive, final answer to the question:

When is it possible to find algebraic formulas for the solutions of polynomial equations?

§11.6   Although Galois' work marked the end of the equation story, it also marked the beginning of the group story. That is how I am treating Galois theory in this book: as a beginning, not an end.

I left the historical thread at the end of §11.3 with Liouville's publication of Galois' papers in 1846. Then, as now, Galois' theory could be taken either as an end or a beginning. Those mathematicians who took notice of Galois' work seem mainly to have adopted the "end" view. *A family of problems dating back for centuries, concerning the solution of polynomial equations, has been wrapped up once and for all. Good! Now let's press forward with the new, promising areas of mathematics: function theory, non-Euclidean geometry, quaternions. . . .*

The first really significant turn toward the future was taken by a mathematician who came, as Abel had, from Norway. This was Ludwig Sylow (pronounced SÜ-lov), who was born in Oslo, still named Christiania at that time, in 1832, the year of Galois' death.

The supply of Norwegian mathematicians being greater than the local demand, Sylow spent most of his working life as a high school teacher in the town of Halden, then called Frederikshald, 50 miles south of Oslo. He did not get a university appointment until he was well into his 60s. All through the years of schoolteaching, though, he kept up his mathematical studies and correspondence.

Sylow was naturally drawn to the work of his fellow countryman Abel on the solvability of equations. Then—this would be in the late 1850s—one of the professors at the University of Christiana showed him Galois' paper, and Sylow began to investigate permutation groups. Bear in mind that abstract group theory, despite Cayley's 1854 paper, was not yet part of the outlook of mathematicians. Group theory was still a theory of permutations, with the only fruitful application being research into the solutions of algebraic equations.

In 1861, Sylow obtained a government scholarship for a year of travel in Europe. He visited Paris and Berlin, attending lectures by several mathematical luminaries of the time, including that Liouville who had resurrected and published Galois' papers 15 years before. On his return to Oslo he gave a course of lectures at the university on permutation groups. This was one of the rare lectures on group theory before the 1870s[112] and is interesting for another reason I shall mention in §13.7.

Sylow's inquiries concerned the structure of permutation groups. I have already mentioned Lagrange's theorem, which imposes a necessary condition for H to be a subgroup of G: The order of H must divide the order of G exactly. Thus a group of order 6 may have subgroups of order 2 and 3, but it may not have subgroups of order 4 or 5.

Sylow's work centers on that word "may." All right, a group of order 6 *may* have subgroups of order 2 and 3. But does it? Can we get some better rules than Lagrange's simple necessary-but-not-sufficient division test for subgroups? Cauchy had already shown that if the order of a group has prime factor p, there is a subgroup of order p. Can this result be improved on?

It certainly can. In a paper published in 1872, Sylow presented three theorems on this topic, still taught to algebra students today as fundamental results in group theory. I shall state only the first here.

> *Sylow's First Theorem.* Suppose G is a group of order n, p is a prime factor of n, and p^k is the highest power of p that divides n exactly. [Example: $n = 24$, $p = 2$, $k = 3$.] Then G has a subgroup of order p^k.

A subgroup of this kind, with order p^k, is called a Sylow *p*-subgroup of G. And it is infallibly the case, at any point in time, that somewhere in the world is a university math department with a rock band calling themselves "Sylow and his *p*-subgroup."

§11.7   The theory of finite groups has subsequently had a long and
fascinating history in its own right. It has also found applications in
numerous practical fields from market research to cosmology. An entire taxonomy of groups has been developed, with groups of different
orders organized into families.

The two groups whose multiplication tables I showed in §10.6
illustrate two of those families. The first was S_3, the group illustrated
by all possible permutations of three objects, which has order 6. The
other was C_6, the group illustrated by the sixth roots of unity, also of
order 6. Both are members of families of groups. The set of all possible permutations of n objects, with the ordinary method for compounding permutations, illustrates the group S_n, the *symmetric group*
of order $n!$ (that is, factorial n). The nth roots of unity, with ordinary
multiplication, illustrate C_n, the *cyclic group* of order n. We have in
fact already spotted a third important family of groups: If you extract
from S_n the normal subgroup of even permutations, that is called the
alternating group of order $\frac{1}{2} n!$—to say it more simply, "of index 2"—
and always denoted by A_n.

Another important family is the *dihedral groups*. "Dihedral"
means "two-faced," in the geometric not the personal sense. Cut a
perfectly square shape from a piece of stiff card. The shape has two
sides—it is dihedral. Label the four corners A, B, C, and D, going in
order like that. Lay the square down on a sheet of paper and pencil
around its outline, to make a square on the paper. Now ask the question: In how many basic, elementary ways, to which all other ways are
equivalent (as a rotation through 720 degrees is equivalent to one
through 360 degrees), can I move this square so that it always ends up
securely and precisely in its outline?

The answer is 8, and I have sketched them in Figure 11-3, showing the effect of each one on a certain starting configuration, represented by the do-nothing identity movement I. There is that identity; there is clockwise rotation through 90, 180, or 270 degrees; and
there are four ways to flip the card over, depending on the axis of

the flip (north-south, east-west, northeast-southwest, or northwest-southeast—these are the four axes of symmetry of the square).

These movements ("transformations" to a mathematician) form a group under the following rule of combination: Apply first one basic movement, then another. Plainly the group has order 8. Its name is D_4, the dihedral group of the square. You might want to try writing up a Cayley table for this group. There is a similar group corresponding to any regular n-sided polygon. That group is called D_n. The case $n = 2$, where the "polygon" is just a line segment AB, has only two members, but for every n from 3 onward, D_n has $2n$ members. So here is another family of groups.

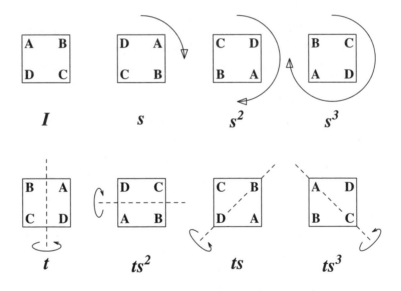

FIGURE 11-3 The eight elements of the dihedral group D_4.

(Note that I have been a little sophisticated with my notation in labeling the members of D_4. Once I have defined the 90-degree rotation as s, for example, and the compounding rule as "do this one, then do that one," the 180-degree rotation is just s^2—do an s, then do another s—and the 270-degree one is s^3 Similarly, I only need define the flip with a north-south axis as t. The one with a northeast-southwest axis is then ts, and so on. When you can use a smallish number of basic elements, like my s and t, to build up a group, those elements are called *generators* of the group. What are the generators of the order-4 groups in Figure 11-2?)

Suppose I did that exercise with a triangle, to get D_3. Wouldn't D_3 have six members? And haven't I already said that there are only two abstract groups of order 6, S_3 and C_6? The answer to this little conundrum is that D_3 is an instance[113] of S_3. If you think about permutations for a moment, you will see why this is so for a triangle but not for a square, nor for any other regular polygon with more than three sides. *Any* permutation of A, B, and C corresponds to a dihedral motion in D_3. This is not so with A, B, C, and D. Permutation (AC) corresponds to the motion ts, but permutation (AB) does not correspond to any motion in D_4 (which, by the way, like S_3, has both normal and plain-vanilla subgroups—you might try identifying them).

I have already mentioned that if p is a prime number, the only group of order p is the cyclic group C_p, modeled by the p-th roots of unity. Well, if p is a prime number greater than 2, there are only two groups of order $2p$. One of them is C_{2p}, the other is D_p.

Looking back to Figure 11-1, which shows the number of groups of each order, you can now very nearly take your understanding up to $n = 11$. The fly in the ointment is $n = 8$, for which there are five different groups. Three of them are cyclic, or built up from cyclic, groups: C_8, $C_4 \times C_2$, and $C_2 \times C_2 \times C_2$. Another one is of course D_4. The fifth is an oddity, the quaternion group, which has the very peculiar property that even though it is not commutative, all its subgroups are normal.

§11.8   I hope by this point you can see the fascination of classifying groups. Perhaps the most historically interesting region of this field of inquiry has been the attempt to classify all *simple* finite groups. A simple group is a group with no normal subgroups. C_p, for prime p, has no proper subgroups at all and so is simple by default. A_n, the group of even permutations of n objects, is simple when n is 5 or more—that, fundamentally, is why the general quintic equation has no algebraic solution. As well as five other families of simple groups, there are 26 "oddities"—one-off simple groups that don't fit into any family. The biggest one of these "oddities" has order

808,017,424,794,512,875,886,459,904,961,710,757,005,754,368,000,000,000.

The classification of all simple finite groups was accomplished in the middle and later 20th century, being finished in 1980. It was one of the great achievements of modern algebra.[114]

Chapter 12

LADY OF THE RINGS

§12.1   A GROUP IS DEFINED RATHER simply by means of just four axioms: closure, associativity, identity, and inverse (§11.4). This very simplicity gives groups a wide scope, just as there is more variety encompassed by the simple description "four-legged creature" than by a lengthier description like "four-legged creature with tusks and trunk." It is the very simplicity of the group definition that allows groups to be applied to things far from the realm of mere numbers. It also gives groups the leeway to possess a complicated and interesting inner structure, with *normal subgroup* as the key concept.

A field (§FT.2) is a more complex object, needing 10 axioms for its definition. Instead of just one basic rule of combination, it has two: addition and multiplication. (Subtraction and division are really just addition and multiplication of inverses: 8 − 3 is the same as 8 + (−3).) This greater complexity keeps the concept of "field" more closely tied to ordinary numbers. It also, paradoxically, restricts their possibilities for having interesting inner structure.

There is another mathematical object much studied by modern algebra: the *ring*. A ring is more complicated than a group but less complicated than a field, so while it is not as wide ranging in its applications as the group, it can roam farther away from ordinary-number applications than a field can. Like a group, a ring can have an interesting inner structure. The key concept here is the *ideal*. I shall explain that at some length in this chapter and the next.

In fact, as often happens with intermediate notions, the ring concept offers mathematicians something of the best from both worlds. It stands, for example, at the center of modern algebraic geometry, source of the deepest and most challenging ideas in modern algebra. It was, however, an unusually long time before the full power of the ring concept came to be appreciated.

It used to be said that ring theory all began with Fermat's Last Theorem. That turns out to have been a mistake. Fermat's Last Theorem is a good hook on which to hang the beginnings of ring theory, though, so that is where I shall start.

§12.2　I mentioned Pierre de Fermat and his last theorem in §2.6. Scribbled in the margin of Fermat's copy of Diophantus's *Arithmetica* around 1637, the theorem asserts that the equation

$$x^n + y^n = z^n$$

has no solutions in positive whole numbers x, y, z, and n when n is greater than 2. Fermat himself, in 1659, sketched a proof for the case $n = 4$, and Gauss provided the complete proof much later. Euler offered a proof for the case $n = 3$ in 1753.

There was then no real progress for half a century until the French mathematician Sophie Germain—second of the three great female mathematicians in this book (giving Hypatia the benefit of the doubt)—proved that Fermat's Last Theorem is true for a large general class of integers x, y, z, and n. It would be too much of a digression to explain which quartets of integers fall into that class. Suffice it to say that the next steps in progress on the theorem built on Sophie Germain's result.

The French mathematician Adrien-Marie Legendre (who was 72 at the time) and the German (despite his French name) Lejeune

Dirichlet separately proved the $n = 5$ case of the theorem in 1825. By this time everyone understood that the only real challenge was to prove the theorem for prime numbers n, so the next target was $n = 7$. Another Frenchman, Gabriel Lamé, solved that one in 1839. At this point, however, the story took a new turn.

§12.3 Just to remind you, the set $\mathbb{Z}$ of integers consists of all the positive whole numbers, all the negative whole numbers, and zero.

$\mathbb{Z}$: $\ldots, -5, -4, -3, -2, -1, 0, 1, 2, 3, 4, 5, 6, 7, \ldots$

(This list heads off to infinity at both left and right.) Now, $\mathbb{Z}$ is not a field. You can add, subtract, and multiply freely, and the answer will still be in $\mathbb{Z}$. However, division works only sometimes. If you divide -12 by 4, the answer is in $\mathbb{Z}$. If you divide -12 by 7, however, the answer is not in $\mathbb{Z}$. Since you can't divide freely without straying outside the bounds of $\mathbb{Z}$, it is not a field.

$\mathbb{Z}$ is, though, sufficiently interesting and important to be worth the attention of mathematicians in its own right. Even if division can't be made to work reliably, you can do a great deal with addition, subtraction, and multiplication. You can, for example, explore issues of factorization and primality (that is, the quality of being a prime number).

Furthermore, there are other kinds of mathematical objects that resemble $\mathbb{Z}$ in allowing addition, subtraction, and multiplication freely but throwing up barriers to division. This is so, for example, with polynomials. Given two polynomials, say $x^5 - x$ and $2x^2 + 3x + 1$, we can add them to get another polynomial (answer: $x^5 + 2x^2 + 2x + 1$), or subtract them (answer: $x^5 - 2x^2 - 4x - 1$), or multiply them (answer: $2x^7 + 3x^6 + x^5 - 2x^3 - 3x^2 - x$). We can't necessarily divide them to get another polynomial, though. In this particular case we certainly can't, though in other cases we sometimes can: $(2x^2 + 3x + 1) \div (x + 1) = (2x + 1)$. Just like integers![115]

This kind of mathematical object, in which the first three pocket calculator rules work reliably, but the fourth doesn't, is called a *ring*.[116] Now you can see what I meant by saying that rings stand between groups and fields in the order of things. "Field" is more tightly defined than "ring," with a full capability for division; "group" is defined more loosely, with only one way to combine elements. Look at the four axioms in §11.4 defining an abstract group. A field needs 10 axioms, a ring only 6.

In the course of some work in number theory, Gauss had discovered a new kind of ring, one involving complex numbers. Gauss did not have the word "ring" available to him—it did not show up until a hundred years later—or even the abstract concept, but a ring is what he discovered nonetheless. This was the ring of what we now call Gaussian integers, complex numbers like $-17 + 22i$, whose real and imaginary parts both belong to $\mathbb{Z}$. You can develop—Gauss *did* develop—an entire integer-like arithmetic with these "complex integers."

This arithmetic is not at all straightforward. Speaking very generally, rings are even less division-friendly than they at first appear to be. You can get some division going in $\mathbb{Z}$ without too much difficulty and develop the theory of prime numbers and factorization familiar from ordinary arithmetic. The fact that the negative numbers are included in $\mathbb{Z}$ adds only a few small and inconsequential wrinkles. Getting a good theory of primes and factorization up and running in other rings is usually much more difficult. It can be done with Gauss's family. It can also be done with a different family that Euler used in his work on the $n = 3$ case of Fermat's Last Theorem, a family that is slightly "bigger" than Gauss's, allowing $\frac{1}{2}\sqrt{3}$ as well as integers. When you get deeper into these kinds of rings, though, ugly things start to happen.

The ugliest thing is that *unique factorization breaks down*. In the ring $\mathbb{Z}$, any integer can be expressed as a product of a unit and a set of primes in just one way. (A unit, in ring theory, is a number that divides into 1. $\mathbb{Z}$ has two units, 1 and -1. Gauss's ring of complex integers has four: $1, -1, i,$ and $-i$.[117]) The integer -28, for instance, factorizes to $-1 \times 2 \times 2 \times 7$. Other than by just rearranging the order of factors, you can't get it to factorize any differently. The ring $\mathbb{Z}$ is blessed with the property of unique factorization.

On the other hand, consider the ring of numbers $a + bi\sqrt{5}$ where a and b are ordinary integers. In this ring, the number 6 can be factorized two ways, as 2×3 and also as $(1 + i\sqrt{5}) \times (1 - i\sqrt{5})$. This is alarming, because all four of those factors are prime numbers in this ring (defined to mean they have no factors except themselves and units). Unique factorization has broken down.

§12.4   This unhappy state of affairs led to the great debacle of 1847. By this point the French Academy was offering a gold medal and a purse of 3,000 francs for a proof of Fermat's Last Theorem. Following his successful assault on the case $n = 7$, Gabriel Lamé announced to a meeting of the Academy, on March 1 of that year, that he was close to completing a general proof of the theorem, a proof for all values of n. He added that his idea for the proof had emerged from a conversation he had had with Joseph Liouville some months before.

When Lamé had finished, Liouville himself stood up and poured cold water on Lamé's method. He pointed out that it was, in the first place, hardly original. In the second place, it depended on unique factorization in certain complex number rings, and this could not be depended on.

Cauchy then took the floor. He supported Lamé, said that Lamé's method might well deliver a proof, and revealed that he himself had been working along the same lines and might soon have a proof of his own.

This meeting of the Academy was naturally followed by some weeks of frantic activity on Fermat's Last Theorem, not only by Lamé and Cauchy but also by others who had been attracted to the cash prize.

Then, 12 weeks later, before either Lamé or Cauchy could announce their completed proofs, Liouville read a letter to the Academy. The letter was from a German mathematician named Eduard Kummer, who had been following the Paris proceedings on the mathematical grapevine. Kummer pointed out that unique factorization would indeed break down under the approaches taken by Lamé and Cauchy, that he himself had proved this three years earlier (though he had published the proof in a very obscure journal), but that the situation could be recovered to some extent by the use of a concept he had published the previous year, the concept of an *ideal factor.*

It used to be said (E. T. Bell says it in *Men of Mathematics*) that Kummer had developed this new concept in the course of work he himself had been doing on Fermat's Last Theorem. Modern scholars, however, believe that Kummer had in fact done no work on the theorem until *after* discovering these ideal factors. Only then, and alerted to the fuss in Paris, had he tackled the theorem.

Within a few weeks after Liouville's reading of his letter, Kummer sent in a paper to the Berlin Academy proving Fermat's Last Theorem for a large class of prime numbers, the so-called regular primes.[118] His proof used the ideal factors he had discovered. This was the last really important advance in the attack on Fermat's Last Theorem for over a century. The theorem was finally proved by Andrew Wiles in 1994.

But what were these "ideal factors"? It is not easy to explain. Historians of mathematics do not usually bother to explain it, in fact, because Kummer's ideal factors were soon superseded by the larger, more general, and more powerful concept of a *ideal,* which is not a number but a ring of numbers. I think this is a bit unfair to Kummer, so here is an outline of his concept.

Kummer was working with *cyclotomic integers,* a concept I shall pause to explain very briefly. The reader may recall the word "cyclotomic" from my primer on roots of unity, §RU.2. When this word shows up in math, you are never far away from the roots of unity. Suppose p is some prime number. What are the pth roots of unity? Well, the number 1 is of course a pth root of unity. The others are scattered evenly round the unit circle in the complex plane, as in Figure RU-1. If we call the first one (proceeding counterclockwise from 1) α, then the others are $\alpha^2, \alpha^3, \alpha^4, \ldots, \alpha^{p-1}$.

A cyclotomic integer is a complex number having the form

$$A + B\alpha + C\alpha^2 + \ldots + K\alpha^{p-1}$$

where all the capital letter coefficients are ordinary integers in $\mathbb{Z}$ and α is a pth root of unity. If p is 3, for example, then the roots of unity are our old pals 1, ω, and ω^2, the latter two being roots of the quadratic equation $1 + \omega + \omega^2 = 0$ (§RU.3). An example of a cyclotomic integer for the case $p = 3$ would be $7 - 15\omega + 2\omega^2$. Note that this is a perfectly ordinary complex number, $\frac{27}{2} - \frac{17}{2}\sqrt{3}\,i$. I have just chosen to write it in terms of 1, ω, and ω^2.

These cyclotomic integers have some weird and wonderful properties. Sticking with the case $p = 3$, for example, from $1 + \omega + \omega^2 = 0$, it follows that for any integer n, $n + n\omega + n\omega^2 = 0$. Since adding zero to a number leaves it unchanged, I can add the left-hand side to $7 - 15\omega + 2\omega^2$, giving $(n + 7) + (n - 15)\omega + (n + 2)\omega^2$, without changing it, a fact you can easily confirm by substituting the actual values of ω and ω^2. Just as $\frac{3}{4}, \frac{6}{8}, \frac{15}{20}, \frac{75}{100}$, and an infinity of other fractions all represent the same rational number, so that second form of my cyclotomic integer, for any value of n at all, will always represent the same cyclotomic integer.

Well, Kummer's work concerned the factorization of these cyclotomic integers. This turns out to be a deep and knotty issue. As you might guess, the problem of unique factorization breaking down soon arises. (Though not *very* soon: It first happens when $p = 23$. This is one reason the theory of ideal factors is hard to illustrate.) This was the particular problem Kummer tackled. He solved it by tightening the ordinary definition of prime number to make it more suitable for cyclotomic integers. Kummer then built up his ideal factors from these "true primes" to get a full theory of factorization for cyclotomic integers.

Out of all this, Kummer was able to prove his great result show-ing that Fermat's Last Theorem is true for regular primes. This was, though, a particular and local application. Before the full power of ring theory could be revealed, a higher level of generalization had to be attained. This higher level was reached by the following genera-tion of mathematicians.

§12.5   Eduard Kummer's 1847 letter to the French Academy had sig-nificance beyond the merely mathematical. Kummer was 37 at the time, working as a professor at the University of Breslau in Prussia. The unification of Germany was still 20 years in the future, but na-tional feeling was strong, and the German people as a people, if not yet a nation, were the great rising force in European culture. Resent-ment of France for the indignities she had inflicted on Germany dur-ing the Napoleonic wars still ran strong after 40 years.[119]

Kummer felt this resentment keenly. His father, a physician in the little town of Sorau, 100 miles southeast of Berlin,[120] had died when Eduard was three years old, from typhus carried into the district by the remnants of Napoleon's Grand Army on its retreat from Russia. As a result, Kummer had grown up in dire poverty. Though he seems to have been a pleasant enough fellow and a gifted and popular teacher, it is hard not to suspect that Kummer must have felt a twinge of satisfaction at showing the French Academy who was boss.

The defeats and humiliations inflicted on the Germans by Napoleon had had a larger consequence, too. They had spurred Prussia, with the lesser German states following close behind, to over-haul her systems of education and of technical and teacher training. The harvest from this, and from the prestige and example of the mighty Gauss, was a fine crop of first-class German mathematicians at midcentury: Dirichlet, Kummer, Helmholtz, Kronecker, Eisenstein, Riemann, Dedekind, Clebsch.

By the time national unification arrived in 1866, Germany could even boast *two* great centers of mathematical excellence, Berlin and

Göttingen, each with its own distinctive style. The Berliners favored purity, density, and rigor;[121] Göttingen mathematics was more imaginative and geometrical—a sort of Rome/Athens contrast. Weierstrass and Riemann exemplify the two styles. Weierstrass, of the Berlin school, could not blow his nose without offering a meticulous eight-page proof of the event's necessity. Riemann, on the other hand, threw out astonishing visions of functions roaming wildly over the complex plane, of curved spaces, and of self-intersecting surfaces, pausing occasionally to drop in a hurried proof when protocol demanded it.

And while this was happening, French mathematics had gone into a decline. That is to speak relatively: A nation that could boast a Liouville, an Hermite, a Bertrand, a Mathieu, and a Jordan was not starving for mathematical talent. Paris's mathematical high glory days were behind her, though. Cauchy died just two years after Gauss, but Cauchy's death marked the end of a great era of mathematical excellence in France, while Gauss's occurred as German mathematics was rising fast.

§12.6   Richard Dedekind was of the best in that midcentury crop of German mathematicians. A serene and self-contained man who cared about nothing very much except mathematics, Dedekind lived a nearly eventless life, most of it as a college teacher in his (and Gauss's) hometown of Brunswick.

Dedekind's contribution to algebra was threefold. First, he gave us the concept of an ideal. Second, he, with Heinrich Weber, opened up the theory of function fields—the theory of which I gave a very brief hint at the end of my primer on fields. (There are more details on this in §13.8.) Third, Dedekind began the process of *axiomatization* of algebra, the definition of algebraic objects as pure abstractions, in the language of set theory. This axiomatic approach, when it reached full maturity a half-century later, became the foundation of the modern algebraic point of view.

The notion of an ideal is not an easy one to communicate to nonmathematicians because illuminating examples do not come easily to hand. An ideal is, first of all, a *subring* of a ring, a ring within a ring. It is, therefore, a family of numbers (or polynomials or whatever other objects the parent ring is composed of), closed under addition, subtraction, and multiplication, imbedded in a larger family of the same type.

An ideal is not just any old subring, though. It has this peculiarity: If you take any one of its elements and multiply this element by one from the larger ring, the result is bound to be within the subring.

Taking $\mathbb{Z}$ as the most familiar ring, an example of an ideal in $\mathbb{Z}$ would be: All multiples of some given number. Suppose we take the number 15, for example. Here is an ideal:

$$\ldots, -60, -45, -30, -15, 0, 15, 30, 45, 60, 75, \ldots$$

The ideal consists of all integers of the form $15m$, where m is any integer whatsoever. Plainly the ideal is closed under addition, subtraction, and multiplication. And, as advertised, if you multiply any element from the ideal, say 30, by any element from the larger ring $\mathbb{Z}$, say 2, the answer is in the ideal: 60.

It would be very nice if I could expand on this by saying: Now take any *two* numbers from $\mathbb{Z}$ and form all linear combinations of them. Take the numbers 15 and 22, for instance, and form all possible numbers $15m + 22n$, where m and n are any integers whatsoever. That would be a more interesting ideal.

Unfortunately, nothing comes of this when working with $\mathbb{Z}$ because $\mathbb{Z}$ is just too simple in its structure. If you let m and n roam freely over $\mathbb{Z}$, $15m + 22n$ takes every possible integer value, as can easily be proved.[122] So the "ideal" you get is just the whole of $\mathbb{Z}$. If, instead of 15 and 22, I had chosen two numbers with a common factor, say 15 and 21, I should just have gotten the ideal generated by 3, their greatest common divisor. So the kind of ideal shown above is the *only* kind in $\mathbb{Z}$, other than $\mathbb{Z}$ itself (and the trivial ideal consisting of just zero). Ideals in $\mathbb{Z}$ are not, in fact, very interesting.

There is a way to say that in formal algebraic language. In any ring the set of multiples of some particular element a is called the *principal ideal* generated by a. A ring like $\mathbb{Z}$, in which every ideal is a principal ideal, is called a *principal ideal ring*. In a ring that is not a principal ideal ring, you can indeed generate ideals by picking two or more elements $a, b, \ldots$ and running all possible combinations of them; $am + bn + \ldots$. That would be called "the ideal generated by a, $b, \ldots$" One way to classify rings, in fact, is by examining the way a ring's ideals are generated. There is, for example, an important type of ring called a Noetherian ring, all of whose ideals are generated by a finite number of elements each.

In rings of complex numbers, ideals become very interesting indeed. Dedekind gave the abstract definition of an ideal—the one I just gave—and then applied it to a wide class of complex-number rings, a much wider class than Kummer had worked with. By doing so he was able to create definitions of "prime," "divisor," "multiple," and "factor" appropriate to any ring at all.

These definitions were expressed in a way more general than any mathematician had attempted before. Dedekind did not completely detach himself from the realm of numbers, but he introduced his mathematical objects—field, ring (he calls it an "order"), ideal, module (a vector space whose scalars are taken from a ring instead of a field)—with defining axioms, as modern algebra textbooks do. Because he did not have the vocabulary of modern set theory to work with, Dedekind's definitions do not look very modern, but he was on the right track.

I shall have more to say about ideals in the next chapter, when I cover algebraic geometry.

§12.7   Once Dedekind's approach had been broadcast and accepted and the concept of an ideal made familiar, it became clear that rings could have interesting internal structure, like groups. That was when ring theory took off. It was still not thought of as ring theory, though.

The people who used it always had some particular application in mind: geometric, analytic, number-theoretic, or most often algebraic—I mean, concerned with polynomials. It was not until the Lady of the Rings came along after World War I that a coherent theory emerged, embracing all these areas and setting them on a firm axiomatic footing.

I shall introduce that lady in the next section. In the 40 years that elapsed between Dedekind's work and hers, the theory was of course pushed forward by numerous mathematicians, including some great ones. The most interesting aspects of those efforts, though, were geometric and so belong in my next chapter. Here I am only going to mention one name from ring theory during that period, for the intrinsic interest of the man and his life.

The name is Emanuel Lasker, and he is mainly remembered not for mathematics but for chess. He was in fact world chess champion for 27 years, 1894–1921—the longest anyone has held that title.

Lasker was born in 1868 in that region of eastern Germany that became part of western Poland after the border rearrangements that followed World War II. His family was Jewish, his father a cantor in the synagogue of their little town, then named Berlinchen, now Barlinek. Lasker learned chess from his older brother and by his early teens was making pocket money by playing chess in the town's coffee houses. He rose fast in the world of chess, winning his first tournament in Berlin at age 20 and becoming world champion at 25 in a series of matches played in North America (New York, Philadelphia, Montreal) against the reigning champion William Steinitz.

Lasker's mathematical education was thorough but was interrupted by his chess activities. After attending the universities of Berlin, Göttingen, and Heidelberg, he studied under David Hilbert at Erlangen University (Germany) from 1900 to 1902 and got his doctorate in that latter year at age 33. His main contribution to ring theory was the rather abstruse notion of a *primary ideal*, somewhat analogous to the powers of primes that you get when you factorize an integer (for example, $6776 = 2^3 \times 7 \times 11^2$). There is a type of ring called a Lasker ring, and a key theorem, the Lasker–Noether theorem, about the structure of Noetherian rings.

Lasker's life ended sadly. He and his wife had settled down to a comfortable retirement in Germany when Hitler came to power in 1933. The Nazis confiscated all of the Laskers' property and drove them penniless out of their homeland. Emanuel Lasker, in his mid-60s, had to take up tournament chess again. He lived in England for two years and then moved to Moscow. When Stalin's Great Purge began swallowing up his Russian friends, he moved to New York, dying there in 1941.[123]

§12.8 Noetherian rings, the Lasker–Noether theorem—obviously there is a person named Noether in this story somewhere. There are in fact two, a lesser and a greater. The lesser was the father, Max Noether. The greater was his daughter Emmy, who brought together all that had been done in the 40 years since Dedekind's ground-breaking work and transformed it into modern ring theory.

Max Noether was a professor of mathematics in the south German town of Erlangen, just north of Nuremberg. Emmy was born there in 1882. Her career must be seen in the context of the German empire in which she grew up, the empire of Bismarck (prime minister and chancellor to 1890) and Wilhelm II (German emperor—Kaiser—from 1888 to 1918). Wilhelmine Germany was an exceptionally misogynist society, even by late 19th-century standards. The German expression *Kinder, Kirche, Küche* (children, church, kitchen), supposedly identifying a woman's proper place in society, is I think known even to people who don't speak German. It was used approvingly of the attitude displayed by Wilhelm II's lumpish consort, the Empress Augusta Victoria, except that on her lips it was supposed to have been uttered as *Kaiser, Kinder, Kirche, Küche.* For further insights into this topic, I recommend Theodor Fontane's 1895 novel *Effi Briest.* Every literate person is familiar with the great French and Russian portrayals of anguished, transgressing 19th-century womanhood,

Flaubert's *Madame Bovary* (1856) and Tolstoy's *Anna Karenina* (1877), but few know the German entry in this field, Fontane's dry, quiet little masterpiece.[124]

Thus, when Emmy Noether decided, around age 18, to take up pure mathematics as a career, she had set herself to climb a steep mountain. This was so even though she had the advantage of a mathematician father, a professor at a prestigious university. In 1900, when Noether made her decision, women were allowed to sit in on university classes only as auditors and only with the professor's permission. Emmy Noether accordingly sat in on math classes at Erlangen, 1900–1902, then at Göttingen, 1903–1904.

By 1907, there had been some modest reforms, and Noether was awarded a doctorate by Erlangen, only the second doctorate in mathematics given to a woman by a German university. The "habilitation" degree, however, the second doctorate that would have allowed her to teach at university level, was still not open to women. For eight years she worked at Erlangen as an unpaid supervisor of doctoral students and occasional lecturer. There was nothing to stop her publishing, and she quickly became known for brilliant work in mathematics.

These were the years following Albert Einstein's unveiling of his special theory of relativity in 1905. Einstein was absorbed in trying to work out his general theory, which aimed to bring gravitation under the scope of his arguments. There were, though, some difficult problems to be overcome. In June and July of 1915 Einstein presented his general theory, unresolved problems and all, in some lectures at Göttingen University. Einstein noted of this event: "To my great joy I succeeded in convincing Hilbert and Klein."

This was an occasion for joy indeed. David Hilbert and Felix Klein were, even at this fairly late point in their respective careers (Hilbert was 53, Klein 66), two giants of mathematics, while Einstein—he was 36—was still not far beyond the *wunderkind* stage. Hilbert and Klein had, of course, followed the development of Einstein's ideas with interest before he came to lecture in 1915. Now "convinced" (convinced, presumably, that Einstein was on the right lines), they gave their attention to the outstanding problems in the general theory. They knew of some work Emmy Noether had done in the relevant areas and invited her to Göttingen.

(Those relevant areas concerned *invariants* in certain *transformations*, ideas I shall clarify in the following pages. The key transformation in relativity theory is the Lorentz transformation, which tells us how coordinates—three of space, one of time—change when we pass from one frame of reference to another. Invariant under this transformation is the "proper time," $x^2 + y^2 + z^2 - c^2t^2$, at least at the infinitesimal level required to make calculus work.)

Noether duly arrived at Göttingen, and within a matter of months she produced a brilliant paper resolving one of the knottier issues in general relativity and providing a theorem still cherished by physicists today. Einstein himself praised the paper. Emmy Noether had arrived.

§12.9   Emmy Noether was now known as a first-class mathematician, but her professional troubles were not yet over. World War I was into its second year—Emmy's younger brother Fritz (another mathematician) was in the army. Göttingen, though liberal by the standards of Wilhelmine universities, still balked at putting a woman on the faculty. David Hilbert, a broad-minded man who judged mathematicians by nothing but their talent, fought valiantly for Noether but without success.

Some of the arguments on both sides have become legendary among mathematicians. The faculty: "What will our soldiers think when they return to the University and find that they are expected to learn at the feet of a woman?" Hilbert: "I do not see that the sex of a candidate is an argument against her admission as a *Privatdozent* [that is, a lecturer supported from fees paid to him by students]. After all, we are a university, not a bathing establishment."[125]

Hilbert's solution to the Noether problem was characteristic: He announced lecture courses in his own name and then allowed Noether to give them.

In the general liberalization of German society following defeat in World War I, however, it at last became possible for a woman to "habilitate" and get a university teaching position, if only of the *Privatdozent* variety, dependent on students paying their lecture fees. Noether duly habilitated in 1919. In 1922, she actually got a salaried position at Göttingen, though she had no tenure, and the meager salary was soon obliterated by hyperinflation.

It was during these early postwar years that Noether gathered up all the work that had been done on rings and turned it into a coherent abstract theory. Her 1921 paper *Idealtheorie in Ringbereichen* ("Ideal Theory in Ring-Fields"—terminology was not yet settled) is considered a landmark in the history of modern algebra, not only laying out key results on the inner structure of commutative rings[126] but providing an approach to the topic that was quickly taken up by other algebraists, the strictly axiomatic approach that became "modern algebra."

Van der Waerden: "At Göttingen I had above all made the acquaintance of Emmy Noether. She had completely redone algebra, much more generally than any study made until then . . . ."

By the early 1930s, Emmy Noether was at the center of a vigorous group of researchers at Göttingen. She still held a low-level position, ill paid and without tenure, but her power as a mathematician was not in doubt. Noether did not at all conform to the standards of femininity current in that time and place, though—nor, it must be said in fairness to her colleagues, any other time and place. She was stocky and plain, with thick glasses and a deep, harsh voice. She wore shapeless clothes and cropped her hair. Her lecturing style was generally described as impenetrable. Her colleagues regarded her with awe and affection nonetheless, though since they were all male, and Kaiser

Wilhelm's Germany was only a dozen or so years in the past, the affection expressed itself in ways that would not be accepted nowadays.

Hence all the disparaging quips, not meant unkindly at the time, that have become part of mathematical folklore. Best known is the reply by her colleague Edmund Landau, when asked if he did not agree that Noether was an instance of a great woman mathematician: "Emmy is certainly a great mathematician; but that she is a woman, I cannot swear." Norbert Wiener described her somewhat more generously as "an energetic and very nearsighted washerwoman whose many students flocked around her like a clutch of ducklings around a kind, motherly hen." Hermann Weyl expressed the common opinion most gently: "The graces did not preside at her cradle." Weyl also tried to take the edge off the appellation *Der Noether* (*der* being the masculine form of the definite article in German): "If we at Göttingen . . . often referred to her as *Der Noether*, it was . . . done with a respectful recognition of her power as a creative thinker who seemed to have broken through the barrier of sex . . . She was a great mathematician, the greatest."

Ill paid and untenured as her position at Göttingen was, Noether lost it when the Nazis came to power in the spring of 1933. Having been once barred from university teaching for being a woman, she was now more decisively barred for being a Jew. The appeals of her Gentile colleagues and ex-colleagues—led, of course, by Hilbert—counted for nothing.

During the Nazi period, there were two common avenues of escape for Jewish or anti-Nazi intellectual talents: to the USSR or to the United States. Emmy's brother Fritz chose the former, taking a job at an institute in Siberia. Emmy went the other way, to a position at Bryn Mawr College in Pennsylvania. Her English was passable, she was only 51, and the college was glad to acquire such a major mathematical talent. Alas, after only two years Emmy Noether died of an embolism following surgery for removal of a uterine tumor. Albert

Einstein wrote her obituary for the *New York Times*,[127] from which the following:

> In the realm of algebra ... which the most gifted mathematicians have [studied] for centuries, she discovered methods of enormous importance. ... [T]here is, fortunately, a minority composed of those who recognize early in their lives that the most beautiful and satisfying experiences open to humankind are not derived from the outside, but are bound up with the development of the individual's own feeling, thinking and acting. The genuine artists, investigators and thinkers have always been persons of this kind. However inconspicuously the life of these individuals runs its course, nonetheless the fruits of their endeavors are the most valuable contributions which one generation can make to its successors.

Math Primer

ALGEBRAIC GEOMETRY

§AG.1   GEOMETRY, AS I SHALL DESCRIBE in the next chapter, has had a crucial influence on modern algebra. My main text will try to track the nature and growth of that influence. Here I only want to introduce a handful of basic ideas about algebraic geometry. In a fine old mathematical tradition, I shall use conic sections as an introduction to this topic.

§AG.2   Conic sections, often just called "conics," are a family of plane curves, the ones you get when your plane intersects a circular cone. (And note that a cone, mathematically considered, does not stop at its apex but extends to infinity in both directions.) In Figure AG-1 the intersecting plane is the page you are reading, which you must imagine to be transparent. The apex of the cone lies behind this plane. In the first picture, the cone's axis is at right angles to the paper, so the intersection is a circle. I then rotate the cone, bringing the further end up. The circle becomes an ellipse. Then, as I tilt the cone up further, one end of the ellipse "goes to infinity," giving a parabola. Tilting beyond that point gives a two-part curve called a hyperbola.[128]

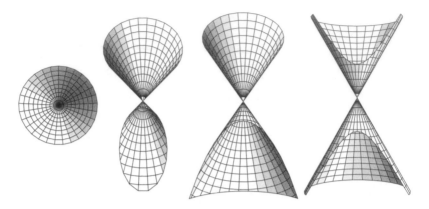

FIGURE AG-1 Conic sections formed by a cone intersecting this page.

That, of course, is all geometry. To get to algebra, we follow Descartes. For the infinity of points that define a conic, Descartes' x and y coordinates satisfy a quadratic equation in x and y, like this one:

$$ax^2 + 2hxy + by^2 + 2gx + 2fy + c = 0$$

(The precise choice of letters for the coefficients there—a, h, b, g, f, c—may seem a little eccentric, but I'll explain that later.) Another way of saying the same thing is: A conic is the *zero set* of some quadratic polynomial in two unknowns. It is the set of points (x, y) that make the polynomial work out to zero.

The ellipse shown in Figure AG-2a, for example, has the equation

$$153x^2 - 192xy + 97y^2 + 120x - 590y + 1600 = 0$$

Now suppose I were to move that ellipse to some other part of the plane and rotate it a little while doing so (Figure AG-2b). What happens to its algebraic equation?

The equation has of course changed. The new equation for my ellipse is

$$369x^2 + 960xy + 1321y^2 + 5388x + 8402y + 18844 = 0$$

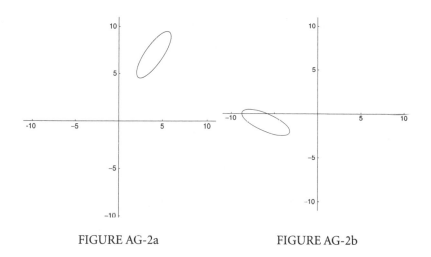

FIGURE AG-2a   FIGURE AG-2b

However, it's still the same conic. Its size and shape haven't changed. It's not larger or smaller, not rounder or skinnier.

So now the following very interesting question arises: What is there in those two algebraic expressions to tell us that they both refer to the same conic? In passing from one equation to the other, what was left unchanged or, as mathematicians say, *invariant*?

The answer is not obvious. All the coefficients changed in size, and in two cases the sign flipped too (negative to positive). There are invariants hidden in there, though. Writing the general case once again as

$$ax^2 + 2hxy + by^2 + 2gx + 2fy + c = 0$$

compute the following things:

$$C = ab - h^2$$

$$\Delta = abc + 2fgh - af^2 - bg^2 - ch^2$$

In the two equations for my ellipse, these work out to $C = 5625$ and 257049, $\Delta = -1265625$ and -390971529. Plainly these aren't invariants either. But look! Calculate Δ^2/C^3 in the two cases. The answer is 9 both times. No matter where we move that ellipse around the plane, how we orient it, what equation we get for it, that calculation from the equation's coefficients will always yield 9. The number Δ^2/C^3 is an *invariant.* In fact, if we take its square root and multiply by π, we have the area of the ellipse, which obviously doesn't change as we slide it around the plane:

$$\text{Area of the ellipse} = \pi \times \sqrt{\Delta^2/C^3}$$

Here's another one. Find the two roots of the quadratic equation

$$t^2 - (a+b)t + C = 0$$

using the coefficients a and b and uppercase C as I calculated it a moment ago. Divide the lesser root by the greater one. Subtract the answer from 1. Take the square root. This number, usually called e (or ε to distinguish it from Euler's number $e = 2.718281828459\ldots$, the base of natural logarithms) measures the *eccentricity* of the ellipse— the degree to which it departs from a perfect circle. If $e = 0$, the ellipse *is* a perfect circle. If $e = 1$, the ellipse is actually a parabola. Plainly, this ought to be an invariant, and it is. If I compute it as I just described for the two equations of my ellipse, e (or ε) comes out to $\frac{2}{3}\sqrt{2}$, which is about 0.94280904—nearer to a parabola than a circle, as you would expect from an ellipse as long and skinny as this one.

What about the actual dimensions of the ellipse? They don't change when we move it around. Shouldn't they be invariants, buried away in those coefficients, too? They are. Referring to the invariant Δ^2/C^3 just for the moment as I, the long axis of the ellipse is

$$2 \times \sqrt{\sqrt{I \div \left(1 - e^2\right)}},$$

while the short one is

$$2 \times \sqrt{\sqrt{I \times \left(1 - e^2\right)}}.$$

If we carry out the arithmetic for the two equations of our ellipse, the long axis comes out to 6 in both cases, the second to 2 in both cases. These numbers are invariants, just as they ought to be.

§AG.3   I have dwelt at some length on this elementary bit of Cartesian geometry because it gives a glimpse of not only the idea of an invariant but also some other key ideas in modern algebra.

Lurking away in that discussion of conics, for instance, is the idea of a *matrix*. For the general equation of a conic as I have given it, the important matrix is

$$\begin{pmatrix} a & h & g \\ h & b & f \\ g & f & c \end{pmatrix},$$

the order of its elements being traditionally remembered by math students via the mnemonic: "All hairy guys [or girls, according to taste] have big feet." From this matrix, or any square matrix, you can extract the determinant, as I described in Chapter 9. The determinant of this matrix is just the number Δ that I defined above.

If you are given the equation of a conic in some Cartesian coordinate system, and you work out the value of Δ, and it turns out that the value is zero, then your conic is a "degenerate" one: a pair of straight lines, or a single straight line, or an isolated point. You might want to confirm that for the isolated point (0,0), equation $x^2 + y^2 = 0$, Δ is indeed zero.

§AG.4   For a glimpse of a different topic, let me return to the issue of the rather peculiar way the coefficients in the general conic's equation are traditionally identified: a, h, b, g, f, c. The omission of d and e is easy to explain: d would be confused with its use in calculus expressions like dy/dx, and e would be confused with Euler's e. But why are the letters all out of order like this? Why not just write the general conic as

$$ax^2 + bxy + cy^2 + fx + gy + h = 0?$$

The answer is that simple Cartesian x,y geometry is not actually the best tool for studying conics. Conics, it turns out, are much more amenable to algebraic treatment if you allow *points at infinity*, which this kind of geometry doesn't. The equation I have given for the general conic comes out of a more sophisticated kind of geometry, one that allows points at infinity.

The phrase "points at infinity" may sound slightly alarming to a nonmathematician. It's just a term of art, though, introduced into geometry[129] to simplify certain matters. If you allow points at infinity, for instance, the awkward business of parallel lines disappears. Instead of

> Any two straight lines in the plane will meet in a single point, unless they are parallel, in which case they won't ever meet at all

you then have

> Any two straight lines in the plane will meet in a single point

... with the understanding that what were formerly thought of as parallel lines meet in a point at infinity. You may not see the importance of this right away, but there is no denying it's a simplification.

Unfortunately, good old Cartesian coordinates in a flat Euclidean plane won't handle this. If you try to write a point at infinity in Cartesian coordinates, all you can come up with is this: (∞,∞). Well, that's one point. It's intuitively clear, however, that if one pair of parallel lines meets at a point at infinity, then another pair, at an angle to the first pair, ought to meet at a *different* point at infinity. In other words, we need more than one point at infinity.

One way to get around this is to replace the ordinary two-coordinate system by a *three*-coordinate system. Instead of identifying a point in the plane by coordinates (x,y), we'll identify it by coordinates (x,y,z). We need to take some slight precautions here to prevent this bursting out into a full-blown three-dimensional geometry, so here's a restraining condition: We'll consider that all (x,y,z) having x, y, and z in the same proportions represent the same point. So for example, $(7,2,5)$, $(14,4,10)$, and $(84,24,60)$ all represent the same point. This isn't such a new idea: Since elementary school you have known that $\frac{3}{4}, \frac{6}{8}, \frac{9}{12}, \frac{30}{40}$, and so on, all represent the same fraction. This restraint squishes the dimensionality back down to 2.

Another way of looking at this new coordinate system is that we have just replaced x and y with $\frac{x}{z}$ and $\frac{y}{z}$. If z is zero, of course, $\frac{x}{z}$ and $\frac{y}{z}$ can't be computed: They are "infinite." The new three-coordinate system sidesteps that little difficulty. We can identify points at infinity by just declaring z to be zero at these points. A point at infinity now looks like this: $(x,y,0)$, with $(2x,2y,0)$, $(3x,3y,0)$, and all other $(kx,ky,0)$ being considered alternate labels for the same point; and there are lots of such points instead of just one.

In fact, they all lie on a line, the line whose equation is $z = 0$. This, you will not be astounded to learn, is called "the line at infinity." There is only one line at infinity, but it's made up of infinitely many different points, each one of which can be nailed down by distinguishing coordinates.

This new kind of geometry, ordinary Cartesian geometry plus a line at infinity, is called *projective geometry*. The new system of coordinates I am describing is a way to bring projective geometry under some kind of arithmetic control. In its purest form, though, projective geometry cares nothing for coordinates. It concerns itself only with those geometric principles that remain true when they are projected. Imagine a geometric figure drawn on a transparency. Hold the transparency at an angle over a flat surface and shine light through it from a point source. The geometric figure is projected onto the flat surface. In doing so, some geometric properties are lost. A circle, for instance, is no longer a circle—it's a conic! Some properties, however, are preserved. I'll say more on this in the next chapter.

§AG.5 What would the equation of a general conic look like in this new coordinate system? Well, let's try just replacing x and y by $\frac{x}{z}$ and $\frac{y}{z}$ in the original equation for a general conic:

$$a\left(\frac{x}{z}\right)^2 + 2h\left(\frac{x}{z}\right)\left(\frac{y}{z}\right) + b\left(\frac{y}{z}\right)^2 + 2g\left(\frac{x}{z}\right) + 2f\left(\frac{y}{z}\right) + c = 0$$

If we multiply both sides of this equation by z^2, we get

$$ax^2 + 2hxy + by^2 + 2gzx + 2fyz + cz^2 = 0$$

Rearranging this slightly,

$$ax^2 + by^2 + cz^2 + 2fyz + 2gzx + 2hxy = 0$$

Now the reason for the order of a, b, c, f, g, and h becomes clear. In this new coordinate system we have an x^2 term, a y^2 term, a z^2 term, a term in yz (that is, xyz with the x left out), a term in zx (xyz with the y left out), and a term in xy (xyz with the z left out). Look at the *symmetry*![130]

From the strictly mathematical point of view, what we have here is not really symmetry, only *homogeneity*. Coordinates of this kind are, in fact, called *homogeneous coordinates*. It's a step in the right direction, though, and shows what a powerful gravitational pull the notion of symmetry has in modern mathematics.

§AG.6 This new coordinate system leads to another topic of key importance in modern math. Once you start to investigate this new arrangement, the arrangement we get when we add points at infinity and a line at infinity, it turns out to be subtly, in fact weirdly, different from the familiar flat Euclidean plane. What, for example, is on the *other side* of the line at infinity? To ask another question: Given a pair of parallel lines that, we have now declared, meet in a point at infinity, which direction should we proceed in along these lines if we want to reach that point? If the parallel lines run east-west, is the point at infinity away to the east or to the west?

Questions of this kind, apparently naïve, turn out, like the questions children sometimes ask, to concern very profound matters. They take us, in fact, into the realm of *topology*.

Topology is generally introduced in pop-math texts as "rubber sheet" geometry. A topologist is interested in those properties of figures that stay unchanged when the figures are deformed by any amount of stretching in any direction, so long as there is no tearing or cutting. Under these rules, for example, the surface of a sphere is equivalent to the surface of a cube, but it is not equivalent to the surface of a doughnut. The surface of a doughnut is, though, equivalent to the surface of a coffee mug with one handle.[131]

Topologically speaking, the good old Euclidean plane is equivalent to the surface of a sphere with a single point missing. (Just imagine the plane "folding up" to cover a sphere that sits on it, but the folding-up never quite making it to the north pole.) Adding that point

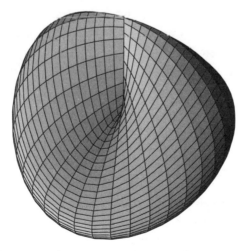

FIGURE AG-3 The projective plane, topologically speaking.

to make a complete sphere does not get us our new arrangement, though.[132] That missing point corresponds to the point at infinity. And our new arrangement—I am going to call it by its proper name, the *projective plane*—doesn't have just one of those; it has infinitely many. So the projective plane is not topologically equivalent to a complete sphere. It is topologically equivalent to a much more peculiar object, a sphere with a crease in it (Figure AG-3).

Note that this object, like the Möbius strip, has only one side. An ant crawling around on it can, provided we permit the one concession of letting him keep going right through the crease, visit every point of the surface, inside and outside. See where a naive question will take you!

§AG.7   *Invariants*, a *matrix* and its *determinant, symmetry, topology*—these are all key ideas in modern algebra. I haven't even finished mining the conics for algebraic issues, in fact.

I spoke of moving that ellipse around the plane. Well, that is one way of looking at what I did. Another would be to think of the ellipse sitting serene and immobile in its plane while the coordinate system moves. (It helps here to think of the x-axis and y-axis, together with a full graph paper grid of squares if you like, printed on a transparency, which is then slid around on the underlying plane.)

Both these approaches offer examples of *transformations*, another very important idea in modern math. These particular transformations, in which only position and orientation change, distances and shapes staying the same, are called *isometries*. They form a study by themselves, of which I shall have more to say in §13.10. And then there are more complicated kinds of transformations: affine transformations (some straight-line stretching and "shearing"—rectangles turning into parallelograms—allowed), and projective transformations (your figure projected as in the last paragraph of §AG.4), topological transformations (stretch and squeeze to your heart's content, but don't cut), Lorentz transformations (in the special theory of relativity), Möbius transformations (in complex variable theory), and many others.

§AG.8   Just one more point arising from those homogeneous coordinates I mentioned, where we use *three* numbers (x,y,z) to identify a point in *two*-dimensional space.

In high school algebra we learn that the Cartesian-coordinate equation of a straight line looks like this: $lx + my + n = 0$. What would that be in homogeneous coordinates?

Well, just writing $\frac{x}{z}$ for x and $\frac{y}{z}$ for y, as I did for the conic, and simplifying, I get this equation:

$$lx + my + nz = 0$$

That's the equation of a straight line in homogeneous coordinates. But look. That means a straight line is determined by the three coefficients (l,m,n), just as a point is determined by its three homogeneous coordinates (x,y,z). More symmetry!

And this raises the question: In a system of homogeneous coordinates, can we build our geometry around lines instead of points? After all, just as a line is an infinity of points (x,y,z) satisfying some linear equation $lx + my + nz = 0$ for fixed coefficients l, m, and n, so a point is an infinity of lines—the lines that all go through that point! Every one of these lines satisfies the equation $lx + my + nz = 0$, but now the point coordinates x, y, and z are fixed while the coefficients l, m, and n roam through an infinity of values, making an infinite "pencil" of lines through the point (x,y,z).

Similarly, instead of thinking of a curve—a conic, for example—as the path traced out by a moving point, we could think of it as traced by a moving line, as I have shown in Figure AG-4.

Can we construct our geometry around this idea? Yes, we can. This "line geometry" was actually worked out by the German mathematician Julius Plücker in 1829, which brings us back to the historical narrative.

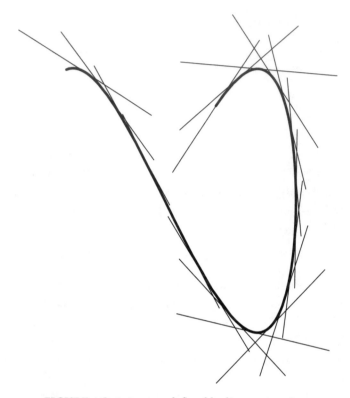

FIGURE AG-4 A curve defined by lines, not points.

Chapter 13

Geometry Makes a Comeback

§13.1   TUCKED AWAY IN SOME glass-fronted cabinet in any university math department, anywhere in the world, is a collection of mathematical models. It usually includes several polyhedra, both convex and stellated (see Figure 13-1), some ruled surfaces done with string, glued-together ping-pong balls illustrating the different methods of sphere packing, a Möbius strip, perhaps a Klein bottle, and other oddities.

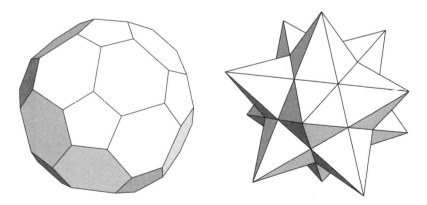

FIGURE 13-1 Polyhedra, convex (left) and stellated (right).

These models are a little faded and dusty nowadays. With the advent of math-graphics software packages such as Maple and Mathematica, which can generate these figures on one's computer screen in a trice, and rotate them for inspection, and transform, deform, and intersect them as desired, it seems absurdly laborious to construct them from wood, card, paper, and string. Making physical models of geometric figures was, though, a favorite pastime for mathematicians and math students through most of the 19th and 20th centuries, and I regret that it seems no longer to be done. I myself spent many happy and instructive hours at it in my adolescence, practically wearing out a copy of H. Martyn Cundy and A. P. Rollett's 1951 classic *Mathematical Models*. My pride and joy was a card model of five cubes inscribed in a dodecahedron, each cube painted a different color.

This interest in visual aids and models for mathematical ideas is one offspring of the rebirth of geometry in the early 19th century. As I mentioned in §10.1, geometry, despite some interesting advances in the 17th century, was overshadowed from the later part of that century onward by the exciting new ideas brought in with calculus. By 1800, geometry was simply not a sexy area of math, and you could be a respectable professional mathematician at that time without ever having studied any geometry beyond the Euclid you would inevitably have covered in school. A generation later this had all changed.

§13.2   The first advance in the 19th-century geometric revolution was made by a Frenchman, Jean-Victor Poncelet, under very trying circumstances. Poncelet, at the age of 24, set off as an officer of engineers with Napoleon's army into Russia. He got to Moscow with the conqueror. In the terrible winter retreat that followed, he was left for dead after the battle of Krasnoy (November 16–17, 1812). Spotting his officer's uniform, a Russian search party carried him off from the battlefield for questioning. Then, a prisoner of war, Poncelet had to walk for five months across the frozen steppe to a prison camp at Saratov on the Volga.

To distract himself from the rigors of imprisonment, Poncelet kept his mind busy by going over the excellent mathematical education he had received at the École Polytechnique. By the time he was allowed to return to France, in September 1814, he had, he tells us, seven manuscript notebooks full of mathematical jottings. These jottings formed the basis for a book, *Treatise on the Projective Properties of Figures*, the foundation text of modern projective geometry.

Poncelet's book was not very algebraic. As a matter of fact, it stood on one side of an argument that was conducted quite passionately through the first half of the 19th century, though it seems quaint now: the argument between *analytic* and *synthetic* geometers. Analytic geometry, descended from Descartes, uses the full power of algebra and the calculus to discover results about geometric figures—systems of straight lines, conics, more complicated curves and surfaces. Synthetic geometry, descended from the Greeks via Pascal, preferred purely logical demonstrations, with as few numbers and as little algebra as possible.

Since its theorems do not mention distances or angles, projective geometry at first looked like a renaissance of the synthetic approach after two centuries of dull Cartesian number crunching. This proved to be a false dawn. Later in the 1820s, German mathematicians—August Möbius, Karl Feuerbach, and Julius Plücker, working independently—brought in homogeneous coordinates, as described in my primer, allowing projective geometry to be thoroughly algebraized.

In addition to Poncelet's founding of modern projective geometry, there was another geometric revolution in the 1820s, for it was in 1829 that Nikolai Lobachevsky published his paper on non-Euclidean geometry in a provincial Russian journal. He then submitted it to the St. Petersburg Academy of Sciences, but they rejected it as being too outrageous. Lobachevsky had argued that the common assumptions of classical geometry—for example, that the three angles in a triangle add up to 180 degrees—might be taken not as *universal truths* about the world but as *optional axioms*. By choosing different axioms, you might be able to get a different geometry, one that didn't look like Euclid's at all—a non-Euclidean geometry.

The young Hungarian mathematician János Bolyai had been working along the same lines. So had Gauss, who in 1824 wrote to a friend: "The assumption that the sum of the three angles of a triangle is less than 180 [degrees] leads to a curious geometry, quite different from ours but thoroughly consistent. . . ." Gauss had in fact been mulling over these ideas for some years. He was, however, a man who cherished the quiet life and avoided controversy, so he never published his thoughts.

Controversial those thoughts were, as Lobachevsky's experience showed. The flat plane (and "flat" space, in its three-dimensional form) geometry of Euclid, with its parallel lines and planes that never meet, its triangles whose angles invariably add up to two right angles, its subtle demonstrations of similarity and congruence, was at this point firmly imbedded in European consciousness. It had been further reinforced by the philosophy of Immanuel Kant, dominant at that time among thinking Europeans. In his 1781 *Critique of Pure Reason*, Kant had argued that Euclidean geometry is (to express it in modern terms) "hard-wired" into the human psyche. We perceive the universe to be Euclidean, Kant argued, because we cannot perceive it otherwise. In that sense, according to Kant, the universe *is* Euclidean, and Euclidean truths lie beyond the scope of logical analysis.[133]

Under these circumstances, the strange new geometries forcing themselves into the consciousness of mathematicians in the 1820s and 1830s were revolutionary—were indeed regarded by good Kantians as subversive. People took their philosophy seriously in those days, when memory of the horrors of the French Revolution and wars were still fresh. What was metaphysically subversive, they thought, might encourage what was socially subversive. If the projective geometry of Poncelet was the first revolution in 19th-century geometry, then the non-Euclidean geometries of Lobachevsky and Bolyai were the second. A third, a fourth, and a fifth were to follow, as we shall see.

§13.3   In the primer preceding this chapter, I mentioned Julius Plücker and his line geometry. Born in 1801—he was a year older than Abel—Plücker had a long career, 43 years (1825–1868) as a university teacher, most of it as a professor at the University of Bonn. His two-volume *Analytic-Geometric Developments* (1828, 1831) was a state-of-the-art account of algebraic geometry in that time, though mostly done with old-fashioned nonhomogeneous coordinates. In the 1830s he worked on higher plane curves, the word "higher" in this context meaning "algebraic, of degree higher than 2"—to put it another way, algebraic curves more difficult than conics.

This work on plane curves was all done from an "analytical" point of view—that is, employing all the resources of algebra, as it existed in the 1830s, and calculus, to deduce laws governing these curves and their properties. Plücker's 1839 book *Theory of Algebraic Curves* dealt definitively with the asymptotes of these curves—that is, with the behavior of the curves near infinity.

Plücker's line geometry came much later, after a 17-year interval (1847–1865) in which he took up physics, occupying the Bonn University chair in that subject. The work on line geometry was in fact unfinished when he died in 1868, and it was left to his young research assistant, Felix Klein, to finish it. I shall have much more to say about Klein a little later.

This interest in curves was a great mathematical growth point in the middle 19th century, nourished by algebra and calculus as well as by geometry. It is an easy interest to acquire, or rather it was in the days before math software came up, the days when you had to work hard, using a lot of computation and a lot of insight, to turn an algebraic equation into a curve on a sheet of graph paper. Who knew, for example, that this rather pedestrian algebraic equation of the fourth degree,

$$4(x^2 + y^2 - 2x)^2 + (x^2 + y^2)(x - 1)(2x - 3) = 0,$$

represents, when you plot y against x, the lovely ampersand shape shown in Figure 13-2? Well, *I* knew, having plotted that curve with

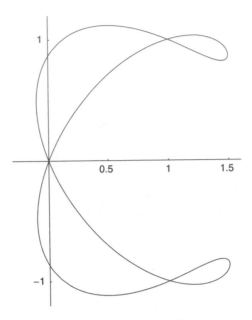

FIGURE 13-2 Ampersand curve.

pencil, graph paper, and slide rule during my youthful obsession with Cundy and Rollett's *Mathematical Models*, which gives full coverage to plane curves as well as three-dimensional figures.

The reader who at this point might be beginning to suspect that the author's adolescence was a social failure would not be very seriously mistaken. In partial defense of my younger self, though, I should like to say that the now-lost practices of careful numerical calculation and graphical plotting offer—offered—peculiar and intense satisfactions. This is not just my opinion, either; it was shared by no less a figure than Carl Friedrich Gauss. Professor Harold Edwards makes this point very well, and quotes Gauss on it, in Section 4.2 of his book *Fermat's Last Theorem*. Professor Edwards:

> Kummer, like all other great mathematicians, was an avid computer, and he was led to his discoveries not by abstract reflection but by

the accumulated experience of dealing with many specific computational examples. The practice of computation is in rather low re-
pute today, and the idea that computation can be *fun* is rarely spoken aloud. Yet Gauss once said that he thought it was superfluous to
publish a complete table of the classification of binary quadratic
forms "because (1) anyone, after a little practice, can easily, without
much expenditure of time, compute for himself a table of any par
ticular determinant, if he should happen to want it . . . (2) because
the work has a certain charm of its own, so that it is a real pleasure
to spend a quarter of an hour in doing it for one's self, and the more
so, because (3) it is very seldom that there is any occasion to do it."
One could also point to instances of Newton and Riemann doing
long computations just for the fun of it. . . . [A]nyone who takes the
time to do the computations [in this chapter of Professor Edwards's
book] should find that they and the theory which Kummer drew
from them are well within his grasp and he may even, though he
need not admit it aloud, find the process enjoyable.

I note, by the way, that according to the *DSB*, Julius Plücker was a
keen maker of mathematical models.

Cundy and Rollett did not pluck my Figure 13-2 out of thin air.
They got it from a little gem of a book titled *Curve Tracing*, by Percival
Frost. I know nothing about Frost other than that he was born in
1817 and became a fellow of King's College, Cambridge, and also of
the Royal Society. His book, first published in 1872, shows the reader
every conceivable method, including some ingenious shortcuts, for
getting from a mathematical expression—Frost does not restrict himself to algebraic expressions—to a plane graph. My own copy of *Curve
Tracing*, a fifth edition from 1960, includes a neat little booklet pasted
to the inside back cover, containing pictures of all the scores of curves
described in the text. The ampersoid curve of my Figure 13-2 is Frost's
Plate VII, Figure 27.

Frost's book in turn looks back to the midcentury algebraic geometers, of whom Plücker was one of the earliest of the big names.
Along the same lines (so to speak) as Frost's *Curve Tracing*, but math-

ematically far deeper, are Irish mathematician George Salmon's four textbooks: *A Treatise on Conic Sections* (1848), *A Treatise on Higher Plane Curves* (1852), *Lessons Introductory to the Modern Higher Algebra* (1859), and *A Treatise on the Analytic Geometry of Three Dimensions* (1862). I have a copy of *Higher Plane Curves* on my desk. It is full of wonderful jargon, mostly extinct now, I am very sorry to say: cissoids, conchoids, and epitrochoids; limaçons and lemniscates; keratoid and ramphoid cusps; acnodes, spinodes, and crunodes; Cayleyans, Hessians, and Steinerians.[134] Homogeneous coordinates had "settled in" by Salmon's time, though he calls them "trilinear" coordinates.[135]

§13.4 Salmon was another "avid computer." In the second edition of his *Higher Algebra* he included an invariant he had worked out for a general curve of the sixth degree. If you look at the invariants for the general conic I gave in my primer, you will believe that this was no mean feat. In fact, it fills 13 pages of Salmon's book.

I mention this as a useful reminder that this midcentury fascination with curves and surfaces was being fed in part from pure algebra and was feeding some results back in turn. The invariants I described in my primer were first conceived in entirely algebraic terms; the geometrical interpretations were secondary.

Arthur Cayley (a close friend of Salmon's, by the way) and J. J. Sylvester were key names in this field from the 1840s on. Most of their work concerned invariants of polynomials, not unlike the ones I illustrated in my primer for the degree-2 polynomial in x and y (or x, y, and z, if we use homogeneous coordinates) that represent a conic. Now, the set of all polynomials in, say, three unknowns x, y, z, with coefficients taken from some field, say the field $\mathbb{C}$ of complex numbers, forms a ring under ordinary addition and multiplication. You can add two polynomials, or subtract them, or multiply them:

$$(2x^2 - 3y^2 + z) \times (y^3z + 4xyz^3)$$
$$= 2x^2y^3z + 8x^3yz^3 - 3y^5z - 12xy^3z^3 + y^3z^2 + 4xyz^4$$

However, you can't reliably divide them. It's a ring. The study of invariants in polynomials is therefore really just a study of the *structure of rings.*

Nobody thought like that in the 19th century. The first glimpses of the larger river—theory of rings—into which the smaller one—theory of invariants—was going to flow, came in the late 1880s, with the work of Paul Gordan and David Hilbert.

Both these men were German, though Gordan was the older by a generation. Born in 1837, from 1874 on he was a professor at Erlangen University, a colleague of Emmy Noether's father. Emmy was in fact Gordan's doctoral student. She is said to have been his only doctoral student, for his style of mathematics—he had studied at Berlin—was highly formal and logical, with much lengthy computation, Roman rather than Greek. By the 1880s, Gordan was the world's leading expert on invariant theory. He had not, however, been able to prove the one key theorem that would have tied up the theory into a neat package. He could prove it in certain special cases but not in all generality.

Enter David Hilbert. Born in Königsberg, Prussia (now Kaliningrad, Russia) in 1862, Hilbert became a *Privatdozent* at that city's university in 1886. Visiting Erlangen in 1888, he met Gordan, and his attention was snagged by that outstanding problem of invariant theory—Gordan's problem, it was called, since Gordan pretty much owned the theory. Hilbert mulled over it for a few months—then solved it!

Hilbert published his proof in December of that year and promptly sent a copy to Cayley at Cambridge. "I think you have found the solution of a great problem," wrote Cayley (who was 68 years old at this point). Gordan was much less enthusiastic. Though Hilbert had not yet held any position at Göttingen, his proof was "very Göttingen" in style: brief, elegant, abstract, and intuitive—Greek, not Roman. This did not suit Gordan's Berlin sensibility. Sniffed he: *Das is nicht Mathematik, das ist Theologie.* ("That's not math, that's theology.") Felix Klein, who *was* at Göttingen, having taken up a professorship there in 1886, was so impressed with the proof that he decided there and then to get Hilbert on his staff as soon as possible.

§13.5   Rather than attempt to describe that proof,[136] I am going to dwell for a moment on Hilbert's slightly later, less sensational, but more accessible result: the Nullstellensatz ("Zero Points Theorem," but always referred to by its German name). The Nullstellensatz introduces the concept of a *variety* and offers an easy connection with geometry.

That connection is an odd one, historically speaking, since Hilbert developed the Nullstellensatz in the context of algebraic number theory, not algebraic geometry. It is a theorem about the structure of commutative rings and properly belongs in ring theory. The algebraic geometers have got their hands on it very firmly by now, though. Open any textbook on algebraic geometry[137] and you will find the Nullstellensatz in the first two or three chapters. I am therefore not going to feel too guilty about offering a geometric interpretation, but I ask the reader to keep in mind that this is really a theorem in pure algebra, in ring theory.

Well, what does it say, this Nullstellensatz? Consider the ring of all polynomials in three unknowns x, y, and z. Just pause to remind yourself that this is indeed a ring: addition, subtraction, and multiplication all work; division works only sometimes. Remind yourself, too, that setting one of those polynomials equal to zero defines some region—usually a curved surface—of three-dimensional space. (I am using ordinary Cartesian coordinates here, not homogeneous ones.) The polynomial $x^2 + y^2 + z^2 - 8$, for example, is equal to zero when (x,y,z) are the coordinates of some point on the surface of a sphere centered on the origin, radius $\sqrt{8}$. Associate that polynomial, in your imagination, with the surface of that sphere.

Now let's consider an ideal in that polynomial ring. An ideal, just to remind you, is a subring, a ring within the ring, having this one extra property: If you multiply any element of the subring by any element of the parent ring, the result is still in the subring.

Consider, for example, all polynomials that look like this: $Ax^2 + Bxy + Cy^2$, where A, B, and C are any polynomials at all in x, y, and z (including the zero polynomial). This is an ideal within the larger ring of all x, y, z polynomials. The polynomial $(x + y + z)(x^2 + y^2)$ is in the ideal; the polynomial $x^3 + y^3 + z^3$ is not.

Now I am going to introduce the key concept of a *variety*, more properly an *algebraic variety*. Geometrically speaking, this is just a generalization of the notion of a curve in two dimensions, or of a surface or "twisted curve" in three dimensions. In fact, a *variety* is the set of zero points of some polynomial, or some family of polynomials.[138]

So the sphere-surface I mentioned above, the points in space at which the polynomial $x^2 + y^2 + z^2 - 8$ is equal to zero, is a variety. So is the intersection of that sphere with the circular cylinder whose equation is $x^2 + y^2 - 4 = 0$. (See Figure 13-3.) That intersection consists of the perimeters of two horizontal circles of radius 2 in three-dimensional space, one at height 2 above the xy coordinate plane, the other at 2 below it. These circles are the variety, the zero-point set, of the polynomial pair $x^2 + y^2 + z^2 - 8$ and $x^2 + y^2 - 4$.

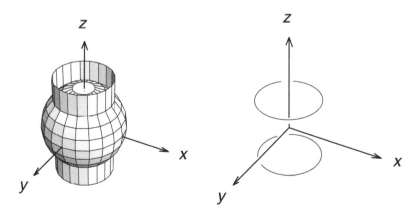

FIGURE 13-3 Two polynomials meet (left) to create a variety (right).

Now, that ideal I described a moment ago is a set of polynomials. It therefore defines a variety—all the points for which *every polynomial in the ideal works out to zero*. What, actually, is that variety, that zero point set? For which set of points do all those polynomials in the ideal work out to zero? The answer is the *z*-axis. The variety defined by that ideal is just that single straight line you get when $x = 0$, $y = 0$, and *z* can be anything you like.

Here is what Hilbert's Nullstellensatz says: If a polynomial equals zero on every point of that variety—in this case, every point of the *z*-axis—then some power of that polynomial is in the ideal. The polynomial $7x - 3y$, for example, is zero on every point of the *z*-axis. It is not in the ideal but its square is: $49x^2 - 42xy + 9y^2$.

I have, of course, oversimplified dramatically there. There need not just be three unknowns *x*, *y*, and *z*; there might be any number. The variety of my example is a particularly simple one . . . and so on. Perhaps the worst of my sins of oversimplification has been the assumption (I didn't say it, but left it "understood") that the coefficients of the *x*, *y*, *z* polynomials making up my ring are real numbers. In fact, they need to be complex numbers—the Nullstellensatz is not generally true for polynomials with real-number coefficients.[139]

In this respect the Nullstellensatz resembles the fundamental theorem of algebra. The two theorems are in fact connected at a deep level, and the Nullstellensatz is sometimes called the fundamental theorem of algebraic geometry. The connection is better expressed by the so-called weak form of the Nullstellensatz: *The variety corresponding to some ideal in a polynomial ring won't be empty (unless the ideal is the whole ring).* The polynomials making up an ideal are bound to have some common zero points—hence the name of the theorem. Compare the FTA, which says that the zero set of a polynomial in one unknown will, likewise, not be empty.

§13.6 Hilbert's statement of the Nullstellensatz came in 1893. That was three further revolutions in geometry on from the midpoint of

the century. Only the last, the fifth, of these revolutions was algebraically inspired, though revolutions three and four had profound ef-
fects on algebra in the 20th century.

Those third and fourth revolutions were both initiated by
Bernhard Riemann, perhaps the most imaginative mathematician
that ever lived. (His dates were 1826–1866.)

In his doctoral thesis at Göttingen in 1851, Riemann presented
his Riemann surfaces, self-intersecting curved sheets that can be used
as replacements for the ordinary complex plane when investigating
certain kinds of functions.

Riemann surfaces arise when we regard a function as *acting on*
the complex plane. The complex number $-2i$, for example, lives on
the negative imaginary axis, south of zero. If you square it, you get
-4, which lives on the negative real axis west of zero. You can imagine
the squaring function having winched $-2i$ around counterclockwise
through 270 degrees to get it to its square at -4.

Bernhard Riemann thought of the squaring function like this.
Take the entire complex plane. Make a straight-line cut from the zero
point out to infinity. Grab the top half of that cut and pull it around
counter-clockwise, using the zero point as a hinge. Stretch it right
around through 360 degrees. Now it's over the stretched sheet, with
the other side of the cut under the sheet. Pass it *through* the sheet
(you have to imagine that the complex plane is not only infinitely
stretchable but also is made of a sort of misty substance that can pass
through itself) and rejoin the original cut. Your mental picture now
looks something like Figure 13-4. That is a sort of picture of the squaring function acting on $\mathbb{C}$.

The power of Riemann surfaces really applies when you look at
them from the *inverse* point of view. Inverse functions are a bit of a
nuisance mathematically. Take the square root function, which is the
inverse of the squaring function. The problem with it is that any nonzero number has *two* square roots. Square root of 4? Answer: 2 . . . or
-2. Both 2 and -2 give result 4 when you square them. There is no

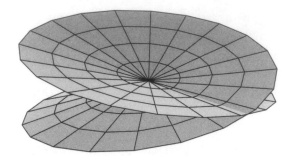

FIGURE 13-4 A Riemann surface corresponding to the squaring function.

getting around this, but Riemann surfaces offer a more sophisticated way to cope with it.[140]

Consider, for example, the fact that the square root of –1 is i . . . or $-i$. Pre-Riemann, a mathematician would have visualized that statement using some image like Figure 13-5.

The Riemann surface shown in Figure 13-4, however, has all the complex numbers stacked up in pairs, one on top of the other (except along the "crease"—but the position of the crease is arbitrary, and if I had available to me the four dimensions I really need for drawing this diagram, I could make it disappear).

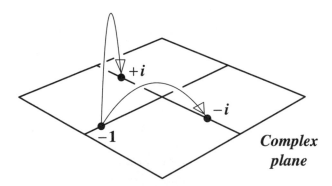

FIGURE 13-5 The number –1 has two square roots
(the pre-Riemann view).

This suggests Figure 13-6 as an alternative way to contemplate the square root function. The Riemann surface that I developed by thinking about the squaring function turns out to be an excellent way to illustrate the *inverse* of that function—the square root function. The number –1 has two square roots, and there they are, on a single line piercing the Riemann surface in two points.

The importance of this first Riemann revolution is that it threw a bridge between the theory of functions, which belongs to the mathematical topic called analysis, and topology—a branch of geometry that had barely gotten off the ground when Riemann came up with all this!

I shall say more about topology in my next chapter. Here I shall only note that the analysis-topology bridge that Riemann created

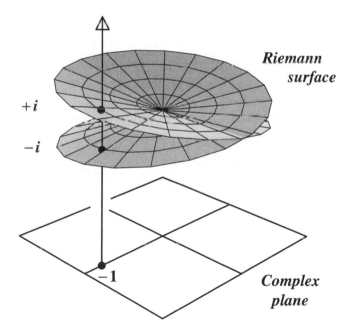

FIGURE 13-6 The number –1 has two square roots
(the post-Riemann view).

opened up function theory to attack by all the sophisticated tools of algebraic geometry and algebraic topology that developed during the 20th century. One of the central theorems here, the Riemann–Roch theorem,[141] relates the analytic properties of a function to the topological properties of the corresponding Riemann surface. In a joint paper published in 1882, Richard Dedekind and Heinrich Weber found a purely algebraic proof of Riemann–Roch, by applying the theory of ideals to Riemann surfaces. That was the result I mentioned in §12.6.

(It may very well be the case, in fact, that Riemann–Roch, in ever more generalized forms, has provided mathematicians of the past 140 years with more work than any other single theorem.)

Not content with having started one revolution in geometry, in 1854 Riemann fired off revolution number four with his stunning habilitation thesis, "On the Hypotheses That Lie at the Foundations of Geometry." Here Riemann created all of modern differential geometry, laying out the mathematics that Albert Einstein would pick up, 60 years later, to use as the framework for his general theory of relativity.

As with Riemann surfaces, the algebraic consequences were indirect. Riemann's thesis provides the primary source for the key 20th-century concept of a *manifold*—a space, of any number of dimensions, that is "locally flat," that is, that can be treated as an ordinary Euclidean space as a first approximation at small scales, just as we regard the curved surface of the Earth as flat for most everyday purposes (see Figure 13-7). This became a key concept in 20th-century algebraic geometry. (The word "manifold"—*Mannigfaltigkeit* in his German—was in fact coined by Riemann, though not in this paper.)

The fifth geometric revolution of the 19th century was the most purely algebraic, though its consequences reached into topology, analysis, and physics. To understand it, we shall have to revisit a topic, and a place. The topic is group theory; the place, the rocky windswept fjords of Norway.

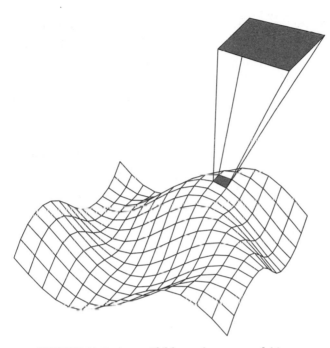

FIGURE 13-7 A manifold may have many folds,
but it is "locally flat" at every point.

§13.7   In §11.6, I mentioned the Norwegian mathematician Ludwig
Sylow and the lectures he gave on group theory at Oslo University in
1862. Among those who attended the lectures was a young fellow
countryman named Sophus Lie, then just 19 years old. At this point
Lie was a full-grown Viking: tall, blond, strong, handsome, and fear-
less. A keen hiker, he was said to be able to cover 50 miles in a day.
That is pretty good over any terrain; over Norwegian terrain, it is
phenomenal. Mathematical folklore says that if it started to rain when
Lie was hiking, he would take off his clothes and stuff them in his
backpack. There is no reference to this, however, in Arild Stubhaug's
meticulous but very respectful biography.[142]

    Although he is now written of as an algebraist, Lie always consid-
ered himself to be a geometer. All his work had a geometrical inspira-

tion. In 1869, on the strength of a paper about projective geometry that he published at his own expense, Lie applied to the university for a grant to visit the mathematical centers of Europe. (It was under one of these grants that Ibsen had made his escape from Norway five years earlier.) Lie left Norway in 1869 and was away for 15 months. He went to Berlin, where he struck up a deep and productive friendship with Felix Klein, who was also visiting that city. Klein had seen Julius Plücker's work on line geometry to publication after Plücker's death; Lie had read the work in Norway and been greatly influenced by Plücker's ideas. Lie went to Göttingen too and then to Paris. In Paris he was joined by Klein again, and the two young men—Lie was 27, Klein had just turned 21—went to lectures given by Camille Jordan.

Though slightly older—he was 32—Jordan was of their own intellectual generation. He had trained as an engineer but was a strong mathematical all-rounder. He had just, in the spring of this year, 1870, published the first book ever written on group theory, the *Treatise on Algebraic Substitutions and Equations*. Jordan's book did not attain the level of generality of Cayley's 1854 papers. He wrote of groups as being groups of permutations and transformations. His coverage of the subject was very comprehensive, though, and the *Treatise* is considered the founding text of modern group theory. How much Lie remembered of Sylow's 1862 lectures we do not know, but it seems certain that it was those weeks in Paris with Jordan that put groups firmly into his head and into Felix Klein's, too.

All this happy and, as we shall see, exceptionally productive mathematical fellowship was rudely interrupted on July 19 when the empire of France declared war on the kingdom of Prussia. Klein, a Prussian, had to leave Paris in haste. In mid-August, Lie too left Paris, heading south for Switzerland on foot, with only a backpack. Thirty miles from Paris he was arrested as a German spy, apparently because he was overheard talking to himself in a language that sounded like German.

Examining the contents of Lie's backpack, the gendarmes found letters and notebooks with German postmarks, filled with cryptic symbols. Lie protested that he was a mathematician. He was ordered to prove it by explaining some of the notes. According to Stubhaug:

> Lie was supposed then to have burst out, saying, "You will never, in all eternity, be able to understand it!" [Words that return an echo from many of us who have tackled Lie Theory . . . —J.D.] But when he realized what danger he was in, he was said nevertheless to have made an effort, and he began thus: "Now then, gentlemen, I want you to think of three axes, perpendicular to each other, the x-axis, the y-axis, and the z-axis . . ." and while he drew figures in the air for them with his finger, they broke into laughter and needed no further proof.

Lie nonetheless had to sit in prison for a month reading Sir Walter Scott's novels in French translation before being allowed to continue to Geneva. When he got back to Christiania in December, he found that he was the 19th-century Norwegian equivalent of a media sensation—the scholar who had been arrested as a spy. The next month he got a research fellowship and lecturing position at the university in Christiania. Shortly afterward, Felix Klein, whose brief service as a medical orderly in the Franco-Prussian War had been curtailed by illness, took up a position as lecturer at Göttingen. Jordan's *Treatise* was being read by mathematicians everywhere. The 1870s, the first great decade of group theory, were under way. So was the fifth revolution in geometry to occur in that amazing century: the "group-ification" of geometry.

§13.8   In Chapter 11, I described several different kinds of groups. Those were all *finite* groups, though. Each had just a finite number of elements. Groups can be infinite as well as finite. The family of integers $\mathbb{Z}$, with ordinary addition as the rule of combination, forms an infinite group.

Geometry is rich with examples of infinite groups. In §11.7, I showed the dihedral group D_4, a group of eight elements that can be illustrated by rotating and flipping a square in such a way that it always occupies the same region of two-dimensional space. That is a finite group, but what if I remove the restriction? What if I allow the square to move around the plane *in any way at all*—sliding to some new position, rotating through any angle at all, flipping over? (See Figure 13-8.) What can be said about *that* family of motions?

What can be said is that it's a group! The operation "moving the square to some new position and orientation" satisfies all the requirements of a group operation. If you do it one way, then follow by doing it another way, the combined result is just as if you had done it

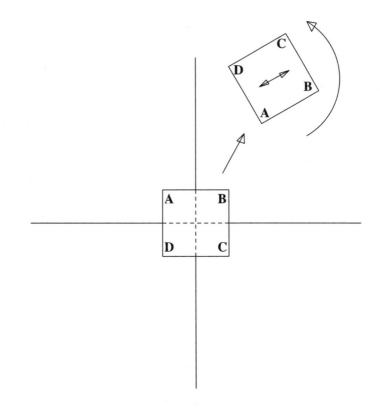

FIGURE 13-8 An isometry.

some third way: $a \times b = c$. The associative law $a \times (b \times c) = (a \times b) \times c$ obviously applies ("do this, then do the result of doing that-and-that . . ."); the do-nothing "motion" will serve as an identity, and every movement can be undone—"has an inverse." It's a group. (Question: Is it commutative?)

Plainly this is a group with infinitely many members. And if you imagine the square drawn on an infinite transparency, sliding and turning as it moves over the "original" plane, you can see that this is *a group of transformations of the entire plane.* Its proper name is, in fact, the group of isometries of the Euclidean plane. The word "isometry" is an important one here. From Greek roots meaning "equal measure," it refers to transformations that "preserve distances." If two points are a distance x apart, they remain x apart under any isometry. There is no stretching or shearing. The distance between any two points is an *invariant* under this group of transformations.

Felix Klein, inspired by his conversations with Lie and Jordan, conceived of a brilliant idea, an idea by means of which the wild jungle growth of geometries that had proliferated in the first 70 years of the 19th century could be unified under a single great organizing principle. The organizing principle should be that a geometry is distinguished by the group of transformations under which its propositions remain true. Two-dimensional Euclidean geometry remains true under the group of plane isometries I just described.[143] And characteristic of that group is some invariant—the invariant in this case being the distance between any two points.

Can we extract some similar group, and characteristic invariant, from projective geometry? From the "hyperbolic geometry" of Lobachevsky and Bolyai? From Riemann's yet more general geometries? We can indeed. The groups here are not easy to describe, but I can at least show you an invariant in projective geometry. Obviously the distance between two points is not preserved in projective geometry. Less obviously, but as illustrated in Figure 13-9, neither is the ratio of two distances between three points: the ratio AC/AB is 2 on the transparency, but 3 on the projection. If you were to take *four*

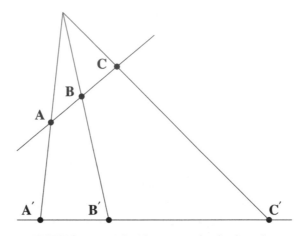

FIGURE 13-9 *AC/AB* not a projective invariant.

points, though, as in Figure 13-10, and compute the ratio of ratios (*AC/AD*) / (*BC/BD*) you would find that it remains the same under projection (in my diagram, it is 5/4), with only some slight, manageable complications when one of the points is projected to infinity. This "cross-ratio" is a projective invariant.

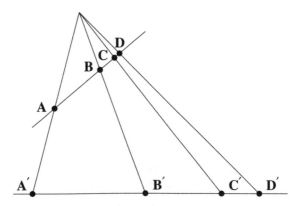

FIGURE 13-10 (*AC/AD*)/(*BC/BD*) *is* a projective invariant.

In the fall of 1872, Klein moved from Göttingen to take up a professorship at Erlangen. It was the custom for a new professor to give an inaugural lecture, a sort of keynote speech for his professorship, laying out the areas of research he intended to encourage. Lie was with Klein in Erlangen for two or three weeks from the beginning of October and helped him work on that speech. In the end, though, Klein did not use this speech for inaugural purposes but published it separately as a paper with the title "Comparative Considerations of Newer Geometric Researches."

This is one of the great mathematical documents of all time, universally known as the Erlangen program.[144] It is not a mathematical paper, in the common sense of one reporting some result or solving some problem. It retains the hortatory quality of an inaugural address, even though Klein did not deliver it at his inauguration. In this program, Klein laid out the idea I sketched above for unifying geometry under the theories of groups and invariants. The program is regarded now, in hindsight, as a bugle call to mathematicians to get busy "group-ifying" their subject.

§13.9   The geometric transformation groups I spoke of in the last section, like the group of isometries in the plane, are not merely infinite; they are *continuous*, their infinity being of the uncountable kind. You can slide that square along an inch, or a thousandth of an inch, or a trillionth of an inch. You can rotate it through 90 degrees, or 90 one-thousands of a degree, or 90 one-trillionths of a degree. There is no limit to how fine you can "cut" these isometries. They can even be "infinitesimal." Translation: In admitting these kinds of transformations into the groupish scheme of things, we have opened the door to let calculus and analysis come into group theory, and vice versa.

Klein himself was not deterred by this. At the end of the Erlangen program, he called for a theory of continuous transformation groups as rigorous and fruitful as the theory of finite groups. ("Of groups of permutations," is what he actually said. The reader must remember that group theory was still far from mature.)

Lie thought this too ambitious, his role in having inspired the Erlangen program notwithstanding. He was now—we are at the end of 1872—a full professor, the Norwegian government having created a chair of mathematics for him. Having little in the way of teaching duties to distract him, Lie absorbed himself in a problem that had gotten his attention. The problem concerned the solving of differential equations, equations in which the unknown quantity to be determined is not a number but a function. An example would be

$$\frac{d^2 y}{dx^2} + 2a\frac{dy}{dx} + p^2 x = 0$$

which can be solved for y in terms of trigonometric and exponential functions of x. Lie had the idea to treat these equations in a way analogous to the way Galois had treated ordinary algebraic equations, but by using these newer continuous groups in place of the finite groups of permutations in Galois theory.

Having gotten fairly deep into this material at the time Klein came out with the Erlangen program, Lie was inclined to think that the whole subject of transformation groups was too large and tangled to permit the kind of rigorous classification Klein suggested. A year later he had changed his mind. A full general theory of continuous groups, of their actions not merely in the plane but on the most general kind of manifold, and of their consequences for the higher calculus was possible. Lie set out to create that theory, and it bears his name to this day.[145]

§13.10   With Klein's announcement of the Erlangen program for the group-ification of geometry, with Lie embarked on his theory of continuous groups, and with Hilbert's discoveries in ring theory around 1890, the picture of 19th-century algebra and algebraic geometry is almost complete. One country has been missing from my account, though, and this is inexcusable in a chapter on algebraic geometry.

The nation of Italy, as we know it today, came into existence in the 1860s following the *Risorgimento* ("re-rising") movement of national consciousness that flourished through the middle decades of the 19th century. Thus relieved from the distractions of getting themselves a nation, the Italians picked up their grand mathematical tradition, populating the later 19th century with some fine scholars: Enrico Betti, Francesco Brioschi, Luigi Cremona, Eugenio Beltrami.

Riemann's influence on these Italian mathematicians was strong. When, in the early 1860s, Riemann's tuberculosis became a hindrance to his work, he traveled to Italy for the warmer air. While in that country he made friends with several mathematicians; it is no coincidence that the student of tensor calculus—the modern development of Riemann's geometry—soon encounters the names of two end century Italian mathematicians, Gregorio Ricci and Tullio Levi-Civita.

The Italians were indeed especially strong in geometry. They took the midcentury approach of investigating curves and surfaces for their own interest—"modern classical geometry," one historian of mathematics[146] called it—and brought it into the 20th century. This was the work of a second cohort of Italian geometers born in the 1860s and 1870s: Corrado Segre, Guido Castelnuovo, Federigo Enriques, and Francesco Severi.

By the time the work of these geometers reached maturity, however, algebraic geometry had ceased to be sexy. This was no mere capricious change in mathematical fashion. By the 1910s, logical and foundational problems had begun to appear in "modern classical geometry," the geometry launched by Poncelet and Plücker a hundred years before. Algebraic geometry was due for an overhaul, via methods presaged by Hilbert and Klein. That overhaul will be one of the topics of my next chapter. Here I only note the achievement of the Italians in keeping algebraic geometry alive while the algebraic tools were prepared for its 20th-century transformation.

The end point of "modern classical geometry" can conveniently be marked, I think, by Julian Lowell Coolidge's 1931 textbook, *A Treatise on Algebraic Plane Curves*. Coolidge, born 1873 in Brookline, Massachusetts,[147] taught at Harvard for most of his life and was chairman of the mathematics department at that noble institution from 1927 until his retirement in 1940. The epigraph of his book reads:

AI GEOMETRI ITALIANI

MORTI, VIVENTI

("To the Italian geometers, dead and living")

Chapter 14

ALGEBRAIC THIS, ALGEBRAIC THAT

§14.1   FROM ABOUT 1870—KLEIN'S Erlangen address forms a convenient milestone—the new understanding of algebra began to be applied all over mathematics. The new mathematical objects (matrices, algebras, groups, varieties, etc.) discovered by 19th-century algebraists began to be used by mathematicians in their work, as tools for solving problems in other areas of math—geometry, topology, number theory, the theory of functions. So far as geometry is concerned, I began to describe some of this spreading algebraicization in Chapter 13. Here I shall extend this coverage to algebraic topology, algebraic number theory, and the further progress of algebraic geometry through the late 19th and early to middle 20th centuries.

§14.2   *Algebraic Topology.* Topology is generally introduced as I described it in §AG.6, as "rubber-sheet geometry." We imagine a two-dimensional surface—the surface of a sphere, say—to be made of some extremely stretchable material. Any other surface this rubber sphere can be transformed into by stretching or squeezing is "the same" as a sphere, so far as the topologist is concerned. There is some lawyering to be added to that, to make topology mathematically precise. You need some rules—which vary slightly in different applications—about cutting, gluing, "pinching off" a finite area into a dimensionless point, or allowing the surface to pass through itself as if the rubber were a kind of mist. The broad, familiar definition will suffice here, though.

Topology did not actually have much to do with algebra until nearly the end of the 19th century. Its early development was, in fact, very slow. The word "topology" first began to be used in the 1840s by the Göttingen mathematician Johann Listing. Many of Listing's ideas seem to have come from Gauss, with whom Listing was close. Gauss, however, never published anything on topology. In 1861, Listing described the one-sided surface we now call the Möbius strip (see Figure 14-1). Möbius wrote the thing up four years later, and for some reason it was his presentation that got mathematicians' attention. It is much too late now to put things right, but I have labeled Figure 14-1 in a way that restores some tiny measure of justice. (If, by the way, you were to take the sphere-with-a-crease of my Figure AG-3 and cut out a small circular patch from it, what was left would be topologically equivalent[148] to a Listing strip.)

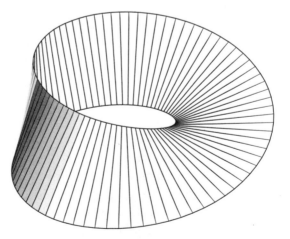

FIGURE 14-1 A Listing strip.

Bernhard Riemann's use of complicated self-intersecting surfaces to aid in the understanding of functions, presented in his 1851 doctoral thesis, was another factor in bringing topological ideas forward. Mulling over these Riemann surfaces, Camille Jordan (see §13.7) came up with the idea of studying surfaces—I am of course using the word "surfaces" here as a more easily visualizable substitute for "spaces"—by seeing what happens to closed loops embedded in them.

Imagine the surface of a sphere, for example. Pick a point on that surface. Starting from that point, walk around in a loop until you arrive back at the point. That path you have traced out: Can it, without having any un-topological violence done to it, and without ever leaving the surface, be shrunk right down to the starting point? *Smoothly and continuously* so shrunk? Yes, it can. So can *any* path on the surface of the sphere.

This is not so on the surface of a torus. Neither of the paths *a* or *b* drawn in Figure 14-2 can be shrunk down to the point *P*, though the path *c* can. So perhaps the study of these paths can indeed tell us something about the topology of a surface.

These ideas were algebraicized in 1895 by a brilliant French mathematician, Henri Poincaré at the École Polytechnique in Paris. Poincaré argued as follows. Consider all possible Jordan loops on a

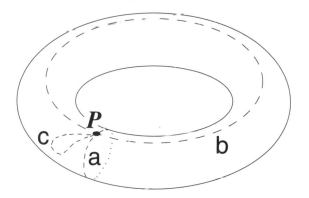

FIGURE 14-2 Loops on a torus.

surface—all those paths you can traverse that begin and end at the same point. Holding that base point fixed, collect all the loops into families, two loops belonging to the same family if one can be smoothly deformed into the other—that is, is topologically equivalent to it. Contemplate these families, however many there might be. Define the following way of compounding two families: Traverse a path from the first family and then a path from the second. (It doesn't matter which paths you choose.)

Here you have a set of elements—the loop-families—and a way to combine two elements together to produce another one. Could it be that these elements, these loop-families, form a group? Poincaré showed that yes, it could, and algebraic topology was born.

It is a short step from this—you need to get rid of dependence on any particular base point for your loops (and they do not, in point of fact, need to be precisely Jordan loops as I have defined them)—to the concept of a *fundamental group* for any surface. The elements of this group are families of paths on the surface; the rule for composition of two path-families is: First traverse a path from one family and then a path from the other.

The fundamental group of a sphere turns out to be the trivial group having only one element. Every loop can be smoothly deformed down all the way to the base point, so there is just one path-family.

The fundamental group of the torus, on the other hand, is a creature known as $C_\infty \times C_\infty$, which looks a bit alarming but really isn't. Take the group elements to be all possible pairs of integers (m, n), with addition serving as the rule of composition: $(m, n) + (p, q) = (m + p, n + q)$ The element (m, n) corresponds to going m times around the a path in Figure 14-5 and then n times round the b path. (If m is negative, you go in the opposite direction. It may help you to follow the argument here if you set off from the base point at an angle, then wind spirally round the torus m times before ending up back at the base point.) These integer pairs, with that simple rule of composition, form the group $C_\infty \times C_\infty$.[149]

I mentioned that the fundamental group for the surface of a sphere is the trivial group with only one member. This fact is itself far from trivial.

It turns out that *any* two-dimensional surface having the trivial one-member group as its fundamental group must be topologically equivalent to a sphere. Now, the familiar two-dimensional sphere-surface embedded in ordinary three-dimensional space has analogs in higher dimensions. There is, for example, a curved three-dimensional space "just like" a sphere but living in four dimensions—a hypersphere, it is sometimes called.[150] Question: Is it also true in four dimensions that any three-dimensional curved space whose fundamental group is the trivial one-member group is topologically equivalent to this hypersphere?

The famous Poincaré conjecture, posed by Henri Poincaré in 1904, asserts that the answer is yes. At the time of this writing (late 2005), the conjecture has neither been proved nor disproved. It is probably true, and in a series of papers made available in 2002 and 2003, the Russian mathematician Grigory Perelman offered a proof that this is so. Mathematicians are still evaluating Perelman's work as I write. Informal reports from these evaluations suggest a growing consensus that Perelman has, in fact, proved the conjecture. The Poincaré conjecture is one of the seven Millennium Prize Problems for the solution of which the Clay Mathematics Institute of Cambridge, Massachusetts, has offered $1 million each. If Perelman's proof is indeed sound, he will therefore become $1 million richer.

When a mathematical theory begins to spawn conjectures, it has truly come alive. Topology came alive with Poincaré's 1895 book, whose title was *Analysis situs*—"The Analysis of Position." That is what topology had mostly been called for the first few decades of its existence, Listing notwithstanding. The use of "topology" to name this subject did not become universal until the 1930s, the credit belonging, I think, to Solomon Lefschetz, concerning whom I shall say more about in just a moment.

§14.3   There is a curious contradiction in Poincaré's having been
the founder of modern topology.

In the minds of mathematicians, topology actually comes in two
flavors, the inspiration for one being geometry, for the other, analysis.
I am using "analysis" here in the mathematical sense, as the branch of
math that deals with functions, with limits, with the differential and
integral calculus, above all with *continuity*. If you look back at my
numerous mentions of *smooth* and *continuous* deformations, you will
grasp the topological connections. In a sense, topology "can't be made
to work" without some underlying notion of smoothness, of continuity, of gliding in infinitely tiny increments from one place to another—some *analytical* way of thinking.

In mathematical jargon, the opposite of *analytical* is *combinatorial.* In combinatorial math we study things that can be tallied off,
one! two! three! . . . , with nothing between the integers. Since there is
no whole number between consecutive integers, there is no smooth
path to take us from one to the next. We just have to leap over a void.
Analytical math is legato, gliding smoothly through continuous
spaces; combinatorial math is staccato, leaping boldly from whole
number to whole number.

Now, topology ought to be the most legato of all mathematical
studies, with its sheets of rubber flexing and stretching smoothly and
continuously. And yet the very first topological invariant that ever
showed up—it measures the number of doughnut-type holes in a
surface and was discovered by Swiss mathematician Simon l'Huilier
in 1813—is a whole number, as well as a hole number . . . . Dimension, which is another topological invariant (you can't, topologically,
turn a shoelace into a pancake, or a pancake into a brick), is also a
whole number. Even those fundamental groups that Poincaré uncovered are not continuous groups, like Sophus Lie's, but countable ones,
as defined in §NP.3. Although they may be infinite, their members
can be tallied: one, two, three, ... You can't do that with the members of a continuous group. So everything of interest in topology
seems to be staccato, not legato.

The paradox is that Poincaré came to topology via analysis, specifically via some problems he was studying in differential equations. Yet his results, and his whole cast of mind in *Analysis situs*, is combinatorial. The analytical approach to topology (the approach nowadays generally called *point-set topology*) had little appeal to him.

This same paradox was even more marked in Poincaré's most important successor in algebraic topology, the Dutch mathematician L. E. J. Brouwer, whose dates were 1881–1966. It was actually Brouwer, in 1910, who proved that dimensionality is a topological invariant. Even more important in modern mathematics is his fixed-point theorem. Formally stated:

Brouwer's Fixed-Point Theorem

Any continuous mapping of an n-ball into itself
has a fixed point.

An n-ball is just the notion of a solid unit disk (all points in the plane no more than one unit distant from the origin) or solid unit sphere (all points in three-dimensional space ditto), generalized to n dimensions. Two-dimensionally, the theorem means that if you send each point of the unit disk to some other point, in a smooth way—that is, points that are very close go to points that are also very close—then some one point will end up where it started.

The fixed-point theorem, together with some straightforward extensions, has many consequences. For instance: Stir the coffee in your cup smoothly and carefully. Some one (at least) point of the coffee—some molecule, near enough—will end up exactly where it started. (Note: Topologically speaking, the coffee in your cup is a 3-ball. By stirring it, you are sending each molecule of coffee from some point X in the 3-ball to some point Y. This is what we mean by "mapping a space into itself.") Even less obvious: Place a sheet of paper on the desk and draw around its outline with a marker. Now scrunch up the sheet, without tearing it, and put the scrunched-up paper inside the outline you marked. It is absolutely certain that one (at least) point of the scrunched-up paper is vertically above the point of the table that point rested on when the paper was flat and you were drawing the outline.[151]

The paradox lurking in Brouwer's topology was that the results he obtained went against the grain of his philosophy. This would not matter very much to an ordinary mathematician, but Brouwer was a *very* philosophical mathematician. He was obsessed by metaphysical, or more precisely *anti*-metaphysical, ideas and with the work of finding a secure philosophical foundation for mathematics.

To this end, he developed a doctrine called *intuitionism*, which sought to root all of math in the human activity of thinking sequential thoughts. A mathematical proposition, said Brouwer, is not true because it corresponds to the higher reality of some Platonic realm beyond our physical senses yet which our minds are somehow able to approach. Nor is it true because it obeys the rules of some game played with linguistic tokens, as the logicists and formalists (Russell, Hilbert) of Brouwer's time were suggesting. It is true because we can *experience* its truth, through having carried out some appropriate mental construction, step by step. The stuff of which math is made (to put it very crudely) is, according to Brouwer, not drawn from some storehouse in a world beyond our senses, nor is it mere language, mere symbols on sheets of paper, manipulated according to arbitrary rules. It is *thought*—a human, biological activity founded ultimately on our intuition of time, which is a part of our human nature.

That is the merest *précis* of intuitionism, which has generated a vast literature. Readers who know their philosophy will detect the influences of Kant and Nietzsche.[152]

Brouwer was not in fact the only begetter of this line of thinking, not by any means. Something like it can be traced all through the modern history of mathematics, back before Kant and at least as far as Descartes. Sir William Rowan Hamilton, the discoverer of quaternions (Chapter 8), can be claimed for intuitionism, I think. His 1835 essay, "On Algebra as the Science of Pure Time," attempts to bring over Kant's idea—mainly based in geometry—of mathematics as founded on "intuition" and "constructions" into algebra.

Later in the 19th century, Leopold Kronecker objected bitterly, on grounds you could fairly call intuitionist *avant la lettre*, to Georg Cantor's introduction of "completed infinities" into set theory. Kronecker argued that uncountable sets like $\mathbb{R}$ do not belong in math—that math can be developed without them, that they bring unwanted and unnecessary metaphysical baggage into the subject, and that mathematics should be rooted in counting, algorithms, and computation.

It is this school of thought that Brouwer brought forward into the 20th century and transmitted to later mathematicians such as the American Errett Bishop (1928–1983). Brouwer's version was called "intuitionism," Bishop's "constructivism." It is as "constructivism" that these ideas are known nowadays, when their leading exponent in the United States is Professor Harold Edwards of the Courant Institute. Professor Edwards's 2004 book, *Essays in Constructive Mathematics*, illustrates the approach very well (as, in fact, do his other books).

Professor Edwards argues that, with easy access to powerful computers, constructivism is now coming into its own and that, once everyone's thinking has made the appropriate adjustments, much of the mathematics done from 1880 onward will come to seem misconceived. I am not qualified to pass judgment on this prediction, but I do personally, as a matter of temperament, find the constructivist approach very attractive and am a big fan of Professor Edwards's writings, as can be seen from the numerous references to them in my text. My remarks on the making of mathematical models and the plotting of curves by hand in Chapter 13 are also very constructivist in flavor.

At any rate, Brouwer's work in algebraic topology, done in his late 20s and early 30s must later have seemed to him philosophically embarrassing. His fellow-countryman B. L. Van der Waerden, who studied under him in Amsterdam 10 years later, said the following in an interview with the *Notices of the American Mathematical Society*:

> Even though his most important research contributions were in topology, Brouwer never gave courses on topology, but always on— and only on—the foundations of intuitionism. It seemed that he was no longer convinced of his results in topology because they were not correct from the point of view of intuitionism, and he judged everything he had done before, his greatest output, false according to his philosophy. He was a very strange person, crazy in love with his philosophy.

§14.4   *Algebraic Number Theory.* This is a phrase not easy to pin down. For one thing, there are objects called algebraic numbers, so that algebraic number theory might, in some context, mean the study of those objects. An algebraic number is a number that is a solution of some polynomial equation in one unknown with whole-number coefficients. Every rational number is algebraic: $\frac{119}{242}$ satisfies the equation $242x - 119 = 0$. Any expression made up of rational numbers, ordinary arithmetic signs, and root signs, is also algebraic: $\sqrt[7]{18 - \sqrt{11}}$ satisfies the equation $x^{14} - 36x^7 + 313 = 0$. It follows from Galois' work, as I described it in §11.5, that a lot of equations of fifth and higher degree have solutions that *can't* be expressed in that way, yet those solutions are algebraic too, by definition. Contrariwise, a lot of numbers are *not* algebraic, π being the best-known example. (Non-algebraic numbers are called *transcendental.* The first proof that π is transcendental was given by Ferdinand von Lindemann in 1882.) There is a large body of theory about these algebraic numbers, built on the foundations laid by Gauss and Kummer, the work I described in §§12.3–4. You can call that algebraic number theory, and mathematicians often do.

Then again, modern algebraic concepts like "group" have proved very useful in tackling problems in traditional, general number theory. Notable among these problems has been the locating of rational points on elliptic curves. That sounds a bit formidable, but in fact the connection here goes straight back to Diophantus and his questions about finding rational-number solutions to polynomial equations like $x^3 = y^2 + x$ (see §2.8). If you were to plot that equation as a graph of y against x, you would have what mathematicians call an elliptic curve. One of Diophantus's solutions would correspond to a point on that curve whose x,y coordinates are rational numbers. Are there any such points? How many are there? How can I locate them? These questions may not sound exciting, but in fact they lead into a fascinating—quite addictive, in fact—area of modern math and to a great unsolved problem: the Birch and Swinnerton-Dyer conjecture.[153]

Yet another topic within the scope of the term algebraic number theory is the discovery, just around the time that Poincaré was launching algebraic topology, of entirely new kinds of numbers, organized—as numbers should be!—in rings and fields. This branch of math was opened up by Kurt Hensel, who was born in Königsberg, David Hilbert's hometown, just 25 days before Hilbert. Hensel studied mathematics under Kronecker in Berlin and then became a professor at Marburg in northwestern Germany, a position he held until his retirement in 1930. His importance to algebra lies in his discovery, around 1897, of p-adic numbers, a brilliant application of algebraic ideas to number theory.

The "p" in p-adic stands for any prime number, so I shall pick the value $p = 5$ for illustration. To begin with, forget about 5-adic *numbers* for a moment while I describe 5-adic *integers*. You probably know that there is a "clock arithmetic" associated with any whole number n greater than 1, an arithmetic in which only the numbers $0, 1, 2, \ldots,$ $n-1$ are used. If we take $n = 12$, for example, we have the ordinary clock face, but with the 12 erased and replaced by 0. If you add 9 to 7 on this clock (that is, if you ask the question: "What time will it be 9 hours after 7 o'clock?"), you get 4. In the usual notation,

$$9 + 7 \equiv 4 \ (\mathrm{mod} \ 12)$$

Okay, write out the powers of 5: 5, 25, 125, 625, 3125, 15625, ... Set up a clock for each of these numbers: a clock with hours from 0 to 4, one with hours from 0 to 24, one with hours from 0 to 124, and so on forever. Pick a number at random from the first clock face, say 3. Now pick a matching number at random from the second clock face. By matching, I mean this: the number has to leave remainder 3 after division by 5. So you can pick from 3, 8, 13, 18, and 23. I'll pick 8. Now pick a matching number at random from the third clock face, a number that will leave remainder 8 after division by 25. So you can pick from 8, 33, 58, 83, and 108. I'll pick 58. Keep going like this . . . forever. You have a sequence of numbers, which I'll pack to-gether into neat parentheses and call x. It looks something like this:

$$x = (3, 8, 58, 183, 2683, \ldots)$$

That is an example of a 5-adic integer. Note "integer," not "num-ber." I'll get to "number" in a minute. Given two 5-adic integers, there is a way to add them together, applying the appropriate clock arith-metic in each position. You can subtract and multiply 5-adic integers, too. You can't necessarily divide, though. This sounds a lot like $\mathbb{Z}$, the system of regular integers, in which you can likewise add, subtract, and multiply but not always divide. It is a *ring*, the ring of 5-adic integers. Its usual symbol is $\mathbb{Z}_5$.

How many 5-adic integers are there? Well, in populating that first position in the sequence, I could choose from five numbers: 0, 1, 2, 3, and 4. Same for the second position, when I could choose from 3, 8, 13, 18, and 23. Same for the third, the fourth, and all the others. So the number of possibilities altogether is $5 \times 5 \times 5 \times \ldots$ forever. To put it crudely and—certainly from the Brouwerian point of view!—im-properly, it is 5^{infinity}.

What other set has this number? Consider all the real numbers between 0 and 1, and represent them in base 5 instead of the usual base 10. Here is the real number $\frac{1}{\pi}$, for example, written in base 5: 0.1243432434442342413234230322004230103420024 .... Obviously, each position of such a "decimal" (I suppose that should be something like "quinquimal") can be populated in five ways. Just like our 5-adic integers! So the number of 5-adic integers is the same as the number of real numbers between 0 and 1.

This is an interesting thing. Here we have created a ring of objects that behave like the integers, yet they are as numerous as the real numbers! In fact, there is a sensible way to define the "distance" between two 5-adic integers, and it turns out that two 5-adic integers can be as close together as you please—entirely unlike the ordinary integers, two of which can never be separated by less than 1. So these 5-adic integers are somewhat like the integers and somewhat like real numbers.

Now, just as with the ring $\mathbb{Z}$ of ordinary integers, you can go ahead and define a "fraction field" $\mathbb{Q}$—the rational numbers—so with $\mathbb{Z}_5$ there is a way to define a fraction field $\mathbb{Q}_5$ in which you can not only add, subtract, and multiply but also divide. That is the field of 5-adic *numbers*.

$\mathbb{Q}_5$, like $\mathbb{Z}_5$, has a foot in each camp. In some respects it behaves like $\mathbb{Q}$, the rational numbers; in others it behaves like $\mathbb{R}$, the real numbers. Like $\mathbb{R}$, for instance, but unlike $\mathbb{Q}$ (or $\mathbb{Z}_5$), it is *complete*. This means that if an infinite sequence of 5-adic numbers closes in on a limit, that limit is also a 5-adic number. This is not so with every field. It is not so with $\mathbb{Q}$, for example. Take this sequence of numbers in $\mathbb{Q}$:

$$\frac{1}{1}, \frac{3}{2}, \frac{7}{5}, \frac{17}{12}, \frac{41}{29}, \frac{99}{70}, \frac{239}{169}, \frac{577}{408}, \cdots$$

Each denominator is the sum of the previous numerator and denominator; each numerator is the sum of the previous numerator and *twice* the denominator. (So $29 = 17 + 12$ and $41 = 17 + 12 + 12$.) All the numbers in that sequence are in $\mathbb{Q}$, but the limit of the sequence is $\sqrt{2}$, which is not in $\mathbb{Q}$.[154] So $\mathbb{Q}$ is not complete.

We can, however, complete $\mathbb{Q}$ by adding the irrational numbers to it: $\mathbb{R}$ is the "completion" of $\mathbb{Q}$. Or rather, $\mathbb{R}$ is *a* completion of $\mathbb{Q}$. The p-adic numbers offer us other ways to complete $\mathbb{Q}$.

Prime numbers, rings and fields, infinite sequences and limits—here we have a mix of notions from number theory, algebra, and analysis. That is the beauty and fascination of p-adic numbers. They were carried forward into mid-20th-century math by Hensel's student and eventual successor in the Marburg chair, Helmut Hasse. Hasse generalized the p-adic numbers by basing them not just on ordinary primes but on "primes" in more general number systems, like those in my primer on field theory and those developed by Gauss and Kummer in their work on factorization of complex and cyclotomic integers.

Hasse went from Marburg to Göttingen in 1934, and thereby hangs a tale. It had been in the previous year, 1933, that the Nazi Party had come to power in Germany. All the Jewish professors at Göttingen were obliged to leave, and many non-Jews who found the Nazis objectionable—I have already mentioned Otto Neugebauer in this context in §1.3—were also driven out or left in protest.

Under the racial classification system of the time, Hasse was not Jewish. Because he had a Jewish ancestor, however, he was not altogether racially "pure" either and so was ineligible for party membership. He seems not to have been at all anti-Semitic, but he was a strong German nationalist and supported Hitler's nonracial policies.

After the resignation of Neugebauer, and then of his successor, Hermann Weyl (another Gentile, though with a part-Jewish wife), Hasse was made head of Göttingen's Mathematical Institute. His motives seem to have been honestly nationalist—to keep German mathematics alive—and he was disliked by the Nazi functionaries he had to deal with, partly for being racially dubious and partly for his intellectual idealism, a quality not much in favor among National Socialists. However, he was dismissed from the university by the British occupation forces following the defeat of Hitler in 1945 and, then in his late 40s, was faced with the necessity to rebuild his professorial career, a thing he did without complaint.

§14.5 *Algebraic Geometry.* I left algebraic geometry, in the form of "modern classical geometry," under the care of early 20th century Italian mathematicians such as Corrado Segre, Guido Castelnuovo, Federigo Enriques, and Francesco Severi. I noted that by the 1910s, geometry in this style was beginning to encounter a crisis of foundations, with awkward conundrums showing up, most of them related to "degenerate cases" of surfaces and spaces, analogous to those degenerate cases of the conics I described in §AG.3. By 1920 these conundrums had become sufficiently serious to stall further progress.

It was clear that algebraic geometry was in need of an overhaul to put the subject on a more solid foundation, as had been done with analysis in the 19th century by a succession of mathematicians from Cauchy to Karl Weierstrass. The overhaul of algebraic geometry in the 1930s and 1940s was likewise a joint effort by several mathematicians. The essence of it was the raising of geometry to a higher level of abstraction.

I have already mentioned Felix Klein's Erlangen program, the idea of tidying up the mess of geometries that had proliferated in the 19th century—projective geometry, non-Euclidean geometry, Riemann's geometry of manifolds ("curved spaces"), geometry done with complex-number coordinates—by using *group* as an organizing principle.

Once mathematicians, following Klein, began to think of the new geometries as a totality, as a single collection of ideas in need of organizing, they began to notice patterns and principles common to all geometries. The idea of making geometry perfectly abstract, without reference to any visualized points or lines in any particular space, took hold, and several mathematicians of the later 19th century—Moritz

Pasch (in Giessen, Germany), Giuseppe Peano (Turin, Italy), Hermann Wiener (Halle, Germany)—attempted this abstraction.

David Hilbert caught this bug in 1892, when he was still a *Privatdozent* at the University of Königsberg. He traveled to Halle with some colleagues to attend a lecture by Hermann Wiener, in which Wiener expounded on his method of abstraction for geometry. Returning to Königsberg, Hilbert's party had to change trains at Berlin. While waiting in the Berlin station, they talked about Wiener's ideas. Hilbert passed the following remark: "One must at all times be able to say 'tables, chairs, and beer mugs' in place of 'points, straight lines, and planes'."[155] (Compare the remarks about algebra made by Peacock, Gregory, and De Morgan in 1830–1850, quoted in §10.1.)

Hilbert did not follow up this memorable apothegm with action for another six years, by which time he was installed as a professor at Göttingen. He then gave a series of lectures during the winter of 1898–1899, in which the traditional geometry of Euclid was derived from a clear, complete set of abstract rules, of axioms, like those I gave for groups in §11.4. The objects referred to by the axioms, said Hilbert, might be any objects at all, but he *chose* to speak of them as points, lines, and planes in order to preserve the clarity of his exposition. These lectures were printed up as a book with the title *The Foundations of Geometry*.

The book was widely read by mathematicians and was very influential. Hilbert's own mathematics subsequently went off in other directions, but he often returned to geometry for brief visits. In the winter of 1920–1921 he gave a series of lectures called "Intuitive Geometry" in which he ranged more widely, but less abstractly, than in the 1898–1899 lectures. This series, too, was printed up as a book, *Geometry and the Imagination*, which still remains popular today.[156]

Hilbert's axiomatic treatment of Euclid's geometry was an inspiration to younger mathematicians. It was some years, however, before the way forward became clear. There were simply too many different viewpoints jostling for attention: Hilbert's axiomatic approach; Klein's group-ification program of 1872 and his reworking of topology in 1895; Hilbert's work on algebraic invariants (the Nullstellen-satz and basis theorems); and, toiling steadily away in the background, the Italian geometers, taking the mid-19th-century approach to the study of curves, surfaces, and manifolds as far as it could be taken.

§14.6   Two names stand out from the crowd in the eventual rework-ing of algebraic geometry: Solomon Lefschetz and Oscar Zariski. Both were Jewish; both were born in the Russian empire, as it stood in the late 19th century.

Lefschetz was the older, born in 1884. Though Moscow was his birthplace, his parents were Turkish citizens, obliged to travel con-stantly on behalf of Lefschetz Senior's business. Solomon was actu-ally raised in France and spoke French as his first language. Of Brouwer's generation, he made his name in algebraic topology, as Brouwer did. In fact, even more remarkably like Brouwer, Lefschetz has a fixed-point theorem named after him. He moved to the United States at age 21 and worked in industrial research labs for five years before getting his Ph.D. in math in 1911. One consequence of this industrial work was that he had both his hands burned off in an elec-trical accident. He wore prostheses for the rest of his life, covered with black leather gloves. When teaching at Princeton (from 1925), he would begin his day by having a graduate student push a piece of chalk into his hand. Energetic, sarcastic, and opinionated, he was something of a character—Sylvia Nasar's book *A Beautiful Mind* has some stories about him. He summed up his own relevance to the history of algebra very vividly: "It was my lot to plant the harpoon of algebraic topology into the body of the whale of algebraic geometry."

Fifteen years younger than Lefschetz, Oscar Zariski was born in 1899. This was a particularly bad time to be born in Russia—to be born Jewish anywhere in the Old World, in fact. The turmoil of World War I, the revolutions of 1917, German occupation, and the subse-quent civil war eventually drove Zariski out of his homeland. In 1920 he went to Rome, where he studied under Guido Castelnuovo, a leader of the Italian school of "modern classical" geometers. By this time Castelnuovo and his colleagues understood that their methods had taken them as far as they could go. Castelnuovo, in his mid-50s at this point, felt that it was time to pass the torch. He urged Zariski to study the topological approach of Lefschetz.

This was at the time in the mid-1920s when Mussolini and his fascists were tightening their grip on Italian public life. Zariski got his doctorate at Rome in 1925. Within a year or two it was plain that Italy was not going to be the refuge from turmoil he had hoped for. Lefschetz was in Princeton now, and Zariski, following Castelnuovo's encouragement, had established a working friendship with him. With Lefschetz's help, in 1927, Zariski got a minor teaching position at Johns Hopkins University in Baltimore. Two years later he joined the faculty of that institution.

Through the late 1920s and early 1930s, Zariski worked away at bringing Lefschetz's modern topological ideas to bear on the "modern classical" geometry he had learned from the Italians. The result of this was a book, *Algebraic Surfaces*, published in 1935.

In the course of writing and researching that book, though, Zariski came to realize that the way forward in algebraic geometry lay not through topology alone but through the axiomatic methods pioneered by Hilbert in *Foundations of Geometry* and applied to abstract algebra by Emmy Noether. (This idea that mathematics had reached a fork in the road was on the minds of many mathematicians in the late 1930s. It is the context for that remark by Hermann Weyl that I quoted in my Introduction.) Beginning in 1937, Oscar Zariski set himself the task of reworking the foundations of algebraic geometry.

Though by this time he had become an American mathematician, Zariski spent the 1945–1946 academic year as a visiting instructor at the University of São Paolo, Brazil. His duties there included giving one lecture course of three hours a week. Only one person attended these lectures, the slightly younger French mathematician André Weil.

Born in 1906, Weil, who was a pacifist as well as being Jewish, had fled from the European war to some teaching positions in the United States before taking a post in São Paolo at the same time as Zariski.[157] He was an established and quite well-known mathematician—Zariski had actually met him at least twice before, at Princeton in 1937 and at Harvard in 1941. The year they spent together in São Paolo was, however, exceptionally productive for both of them.

Weil, like Zariski, had gotten the idea of reworking algebraic geometry using the abstract algebra of Hilbert and Emmy Noether. In particular, he worked to generalize the theory of algebraic curves, surfaces, and varieties so that its results would be valid over *any* base field—not just the familiar $\mathbb{R}$ and (by this time) $\mathbb{C}$ of the "modern classical" algebraic geometers but also, for example, the finite number fields I described in my primer on field theory. That opened up a connection with the prime numbers and with number theory in general, and Weil's work was fundamental to the algebraicization of modern number theory. Without it, Andrew Wiles's 1994 proof of Fermat's last theorem would have been impossible.

The various streams of thought that arose in the 19th century were now about to flow together into a new understanding of geometry, one based in abstract algebra and incorporating themes from topology, analysis, "modern classical" ideas about curves and surfaces, and even number theory. Hilbert's beer mugs and Emmy Noether's rings, Plücker's lines and Lie's groups, Riemann's manifolds and Hensel's fields, were all brought together under a single unified conception of algebraic geometry. That was one of the grand achievements of 20th-century algebra, though by no means the only one— nor the least controversial.

Chapter 15

FROM UNIVERSAL ARITHMETIC TO UNIVERSAL ALGEBRA

§15.1   AS A GLIMPSE OF ACADEMIC WORK in algebra over recent decades, consider the following fragments extracted from a list of awards of the Frank Nelson Cole Prize in Algebra, given by the American Mathematical Society. (The full list is available on the Internet.)

> *1960:* To Serge Lang for his paper "Unramified class field theory over function fields in several variables," and to Maxwell A. Rosenlicht for his papers on generalized Jacobian varieties . . . *1965:* To Walter Feit and John G. Thompson for their joint paper "Solvability of groups of odd order." . . . *2000:* To Andrei Suslin for his work on motivitic cohomology . . . *2003:* To Hiraku Nakajima for his work in representation theory and geometry.

Surveying that list, the reader might be excused for thinking that I have skimped on my coverage of algebra in this book. Jacobian varieties? Unramified class field theory? Motivitic cohomology? What *is* this stuff?

Well, it is modern algebra, built up around key concepts such as *group, algebra, variety, matrix,* all of which I hope I *have* given some fair account of. Even some of the unexplained terms are only a step or two removed from these basic 19th-century ideas. *Representation,* for example, refers to the study of groups and algebras by means of families of matrices that model them, a thing I touched on in §9.6. Class field theory is a very generalized, modernized approach to the problems raised by non-unique factorization, the problems that so vexed Cauchy and Lamé back in §12.4. *Solvability* refers to matters of group structure, harking all the way back to the solvability of equations . . . and so on.

It is a fact, though, that algebra has become very abstruse and that topics such as motivitic cohomology simply cannot be made accessible to a reader who does not have a math degree—nor even, I think, to a reader who has one unless he has specialized in the right area.[158] Algebra has also become very *large,* embracing a diversity of topics—13 out of the 63 subject headings in the current (2000) classification system used by the American Mathematical Society.[159]

At this point, therefore, I am going to exercise author's privilege and just offer three sketches of topics and personalities from the last few decades, without any claim that this will give a complete picture of recent algebra. The first sketch, §§15.2–15.5, will deal with category theory; the second, §§15.6–15.9, with the life and work of Alexander Grothendieck; and the third, §§15.10–15.11, with applications of modern algebra to physics. I shall postpone discussion of motivitic cohomology to some future book . . . .

§15.2   One of the most popular textbooks for math undergraduates in the later 20th century was Birkhoff and Mac Lane's *A Survey of Modern Algebra*. First published in 1941, it brought all the key concepts of mid-20th-century algebra together in a clear, connected presentation, with hundreds of exercises for students to sharpen their wits on. Numbers, polynomials, groups, rings, fields, vector spaces, matrices, and determinants—here it all was. I myself learned algebra from Birkhoff and Mac Lane, and I am sure my book shows the influ-

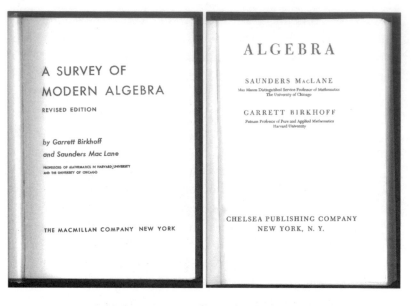

FIGURE 15-1 Birkhoff and Mac Lane (1941),
and Mac Lane and Birkhoff (1967).

ence of theirs. (It actually shows more than that; I have borrowed a couple of their exercises to help make my points.)

In 1967, a completely new edition of the book appeared. The title was changed to just *Algebra*. The authors were listed in reverse order: Mac Lane and Birkhoff. Most significantly, the presentation was changed. The fourth chapter was entirely new. Titled "Universal Constructions," it dealt with functors, categories, morphisms, and posets—terms that do not appear at all in the 1941 *Survey*. A lengthy (39 pages) appendix on "Affine and Projective Spaces" was added.

That an undergraduate math text should have needed such extensive revision just 26 years after first publication is a bit surprising. What had happened? Where did these new mathematical objects, since presumably that is what they are—these functors and posets— where did they suddenly spring from?

Garrett Birkhoff (1911–1996) and Saunders Mac Lane (1909–2005) were both instructors at Harvard in the late 1930s. Birkhoff's father, George Birkhoff, was himself a professor of mathematics at that university from 1912 until his death in 1944. It was the elder Birkhoff that Albert Einstein famously described as "one of the world's great anti-Semites," though Birkhoff Senior's prejudices, while real, seem not to have been extraordinary in that time and place.[160] The younger Birkhoff was appointed instructor at Harvard in 1936. Mac Lane, the son of a Congregational minister in Connecticut, taught at Harvard from 1934 to 1936 and was appointed assistant professor there in 1938.

As university teachers of algebra, both men were strongly influenced by a book published in German in 1930. This was B. L. van der Waerden's *Modern Algebra*, the first really clear exposition, at a high mathematical level, of the entirely abstract axiomatic approach to the new mathematical objects that had emerged in the 19th century. Van der Waerden subtitled the book *Using the lectures of E. Artin and E. Noether*. "E. Noether" is of course the Emmy Noether of my §12.9. Emil Artin was a brilliant algebraist at Hamburg University until the Nazis came to power, after which he taught at various universities in the United States. The original idea, in fact, had been that Artin and van der Waerden should write the book jointly, but Artin backed out of the project under pressure of research work. Van der Waerden's book brought together all the new mathematical objects—groups, rings, fields, vector spaces—and gave them the abstract axiomatic treatment, as Hilbert, Noether, and Artin had developed it.

Van der Waerden passed on this way of thinking to the mathematical community at large through his 1930 book. Birkhoff and Mac Lane made it accessible to undergraduates with their 1941 book. From that point on, the term "modern algebra" had a distinct meaning in the minds of mathematicians and their students. The essence of that meaning was an approach to algebra that was perfectly abstract and carefully axiomatic, all expressed in the language of set theory, like the definition of "group" that I gave in §11.4.

Was this the last word in abstraction, the end point of the line of thought first voiced by George Peacock in 1830 (§10.1)? By no means!

§15.3   In 1940, while *A Survey of Modern Algebra* was being prepared for the press, Saunders Mac Lane attended a conference on algebraic topology at the University of Michigan. There he encountered a young Polish topologist, Samuel Eilenberg, who had moved to the United States the previous year and whose published papers Mac Lane was already familiar with. They struck up a friendship and in 1942 produced a joint paper on algebraic topology. The paper's title was "Group Extensions and Homology." It deals with homology, about which I ought to say a few words.

In §14.2, I described the *fundamental group* of a manifold in terms of families of loops—closed paths—embedded in the manifold. This fundamental group of path-families is one instance of a *homotopy group*. It is possible to work out other homotopy groups associated with a manifold, by generalizing from those paths—those one-dimensional loops, each of which is topologically equivalent to a circle—to two-, three-, or more-dimensional "hyper-loops," equivalent to spheres, hyperspheres, and so on.

These homotopy groups are interesting and important, but for giving us information about a manifold, they have certain drawbacks. They are, from a mathematical point of view, unwieldy.[161]

Poincaré uncovered a quite different family of groups that can be associated with any manifold. These are the *homology groups*. The most straightforward way to construct homology groups for a manifold is to replace the manifold by an approximate one made up entirely of *simplexes*. You can get the idea by imagining the surface of a sphere deformed into the surface of a tetrahedron (that is, a pyramid on a triangular base; see Figure 15-2). You now have a figure made up of zero-dimensional vertices, one-dimensional edges, and two-dimensional triangular faces. By studying the possible ways to traverse these vertices, edges, and faces, by way of paths that are allowed to

FIGURE 15-2 *Left to right:* A 0-simplex (or point),
a 1-simplex (line segment), a 2 simplex (triangle),
a 3-simplex (tetrahedron), and a 4-simplex (pentatope).[162]

"cancel out" when they traverse in opposite directions (see Figure 15-3), you can extract a family of groups, usually denoted H_0, H_1, and H_2. These are the homology groups and are collectively known as the *homology* of the surface. Furthermore, there is a way do this whole process *in reverse*, treating the vertices as faces, the faces as vertices, and the edges as (differently organized) edges.[163] You then get a different family of groups, collectively known as the *cohomology*.

Something similar can be done with any kind of manifold in any number of dimensions. Now, a triangle is the simplest possible plane polygon enclosing any area at all. To a mathematician, it is a *2-simplex*. A 3-simplex is a tetrahedron—a triangular pyramid, with four vertices and four triangular faces. A 4-simplex is the equivalent thing in four dimensions, having five vertices and five tetrahedral "faces" (see Figure 15-2). For the sake of completion, we can call a line segment a 1-simplex and a single isolated point a 0-simplex.

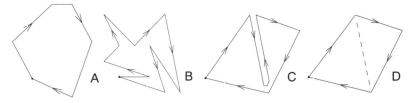

FIGURE 15-3 Paths A and B are *homotopically* equivalent;
paths C and D are *homologically* equivalent.

Any manifold can be "triangulated" like this into simplexes, though if the manifold has holes going through it, like a torus, you will need to glue several simplexes together to make a "simplicial complex." Once you have triangulated your manifold in this way, you can work out the homology—the *simplicial homology*—of the triangulated manifold. This homology, and the corresponding cohomology, carry useful information about the manifold. Furthermore, the groups that comprise the homology are easier to deal with than are homotopy groups. (I note, in passing, the similarity of this procedure to the actual triangulation carried out by mapmakers when surveying a landscape. See §15.9 below.)

We have a glimpse here of a key notion in late 20th-century algebra, the notion of *attaching algebraic objects to a manifold*. The objects I have mentioned are groups that arise when we conduct topological investigations. In homology theory, however, we can also attach vector spaces and modules (see §12.6, penultimate paragraph) to a manifold. This opens up rich new territory in algebraic topology and algebraic geometry. The most famous explorers of that territory were the French mathematicians Leray, Serre, and Grothendieck, about whom I shall say more later.

Well, this was the background to Eilenberg and Mac Lane's 1942 paper. Under the inspiration of the "modern algebra" that was in the air everywhere at that time, and about which Mac Lane had just (with Birkhoff) written a brilliant book, they dealt with the topic very abstractly—their treatment is far, far removed in abstraction from the tiny sketch I gave, from triangles and tetrahedrons. Yet in writing the paper it occurred to both of them that a still higher level of abstraction was possible.

Three years later they attained that higher level, in another joint paper titled "General Theory of Natural Equivalences." This is the paper that launched category theory on the world.

Category theory, which I am going to describe in just a moment, emerged from homology theory in a natural way. In the 40 years since homology groups had been identified in their original topological context, they had been seen to have deep connections with other branches of algebra, in particular with Hilbert's work on invariants in polynomial rings that I sketched in §§13.4-5. Via the connection Riemann had established between function theory and topology (§13.6), they also had relevance to analysis—to the higher calculus, the study of functions and families of functions. A little later, in the 1950s, all this blossomed into a field of study called *homological algebra*. Eilenberg co-authored the first book on homological algebra (with the great algebraic topologist Henri Cartan) in 1955. The level of generality here was so high that category theory was a natural parallel development.

§15.4   The general line of thought underlying category theory is as follows.

Algebraic objects such as groups, rings, fields, sets, vector spaces, and algebras are made up of (a) elements (for example, numbers, permutations, rotations) and (b) a method or methods of combining elements (for example, addition, or addition and multiplication, or compounding of permutations). These objects tend to reveal their structure most clearly when we find (c) ways to transform—to "map"—one of them into another, or into itself. (Recall, for example, how in my primer on vector spaces I mapped a vector space into its own scalar field. Recall also my thumbnail sketch of Galois theory and its central concern with permuting—mapping—a solution field into itself while leaving the coefficient field unaltered.)

Although these are different kinds of objects, with different possibilities for mapping, there are broad similarities of structure and method across the (a), the (b), and the (c) in all cases. Take, for instance, the relationship of an ideal to its parent ring (§12.6), and the relationship of a normal subgroup to its parent group (§11.5). There is something naggingly similar about the two relationships. Is it possible to extract some general principles, *a general theory of algebraic structures*, so that *all* these objects, and any others we might come up with in the future, can be brought under a single set of super-axioms? A sort of universal algebra?[164]

Eilenberg and Mac Lane gave the answer: Yes, it is possible. Wrap up some family of mathematical objects—groups, perhaps, or vector spaces—with some "well-behaved" family of mappings among them. This is a *category*, and the mappings in it are called *morphisms*. You can now go ahead (with care) and set up hyper-mappings from one category (including all its morphisms) to another. This kind of hyper-mapping is called a *functor*.

By way of illustration, look back to my discussion of *p*-adic numbers in § 14.4. I constructed a system of 5-adic integers. Then I said, rather glibly: "[J]ust as with the ring $\mathbb{Z}$ of ordinary integers, you can go ahead and define a 'fraction field' $\mathbb{Q}$—the rational numbers—so with $\mathbb{Z}_5$ there is a way to define a fraction field $\mathbb{Q}_5$ in which you can not only add, subtract, and multiply but also divide." Hidden in that little bit of sleight of hand is the category-theoretic notion of a *functor*.

$\mathbb{Z}$ is, in fact, slightly more than a mere ring. It is a rather particular kind of ring, the kind called an *integral domain*—that is, a ring in which multiplication is commutative (it doesn't have to be for a ring) and has an identity element "1" for multiplication (a thing rings don't have to have) and permits $a \times b = 0$ only if a, b, or both, is/are zero (not an essential condition for a ring). The way to get from $\mathbb{Z}$ to $\mathbb{Q}$ is to create a fraction field from an integral domain—a thing that can generally be done, from any integral domain, because it is possible to construct a functor from the category of integral domains with the mappings between them (well, at any rate, a subset of those mappings), to the category of fields and (a similar subset of) *their* mappings.

Though I really don't want to get too deep into this, I cannot forbear a mention of my favorite functor: the *forgetful functor*. That is the one that maps from a category of algebraic objects—groups, say— into the category of bland unvarnished sets, "forgetting" all the structure that exists in the original objects.

§15.5   Can useful math really be done at such a very high level of abstraction? It depends who you ask. To this day (2006), category theory is still controversial. Many professional mathematicians— rather especially, I think, in English-speaking countries—frown and shake their heads when you mention category theory. Only a minority of undergraduate courses teach it. None of the words "category," "morphism," or "functor" appears anywhere in the 600-odd pages of Michael Artin's magisterial 1991 undergraduate textbook *Algebra*.

When I myself was a math undergraduate in the mid-1960s, the opinion most commonly heard was that while category theory might be a handy way to organize existing knowledge, it was at too high a level of abstraction to generate any new understanding (though I should say that this was in England, where category theory, its American origins notwithstanding, dwelt in the odium of being suspiciously *continental*).

Saunders Mac Lane, at any rate, was very much taken with his and Eilenberg's creation. When, in the mid-1960s, the time came to put out a revised edition of the *Survey of Modern Algebra*, he reworked the entire book to give it a category-theoretic slant. Others have followed him, and if category theory is still not universally accepted, certainly not for the undergraduate teaching of algebra, it has a large and vigorous cheering section in the math world. Adherents are sufficiently confident to make fun of their pet. F. William Lawvere opens his book on the application of categories to set theory by saying: "First, we deprive the object of nearly all content. . . ." Robin Gandy, in *The New Fontana Dictionary of Modern Thought*, wrote: "Those who like to work on particular, concrete problems refer to [category theory] as 'general abstract nonsense.'"[165]

Promoters of category theory make very large claims, some of them going beyond math into philosophy. In fact, category theory from the very beginning carried some self-consciously philosophical flavor. The word "category" was taken from Aristotle and Kant, while "functor" was borrowed from the German philosopher

Rudolf Carnap, who coined it in his 1934 treatise *The Logical Syntax of Language*.

The philosophical connotations of category theory lie beyond my scope, though I shall make some very general comments about them at the end of this chapter. Certainly, though, there have been working mathematicians who have used the theory to obtain significant results. There has, for example, been Alexander Grothendieck.

§15.6   Grothendieck is the most colorful and controversial character in the recent history of algebra. There is a large and growing literature about his life, by now probably exceeding what has been written about his mathematical work. The most accessible and informative account of both life and work so far produced in English is Allyn Jackson's "As If Summoned from the Void: The Life of Alexandre Grothendieck," published in two parts in the October and November 2004 issues of the *Notices of the American Mathematical Society*, the two parts together running to about 28,000 words. There are also numerous Web sites given over to discussions of Grothendieck. A good starting point for English-speaking readers (though it also contains much in French and German) is *www.grothendieck-circle.org*, which includes both parts of Allyn Jackson's aforementioned biographical article.

Grothendieck's story is compelling because it conforms to archetypes about certain fascinating "outsider" personalities: the Holy Fool, the Mad Genius, the Contemplative Who Withdraws from the World.

To take the genius first: Grothendieck's years of glory were 1958–1970. The first of those years marked the founding in Paris of the Institut des Hautes Études Scientifiques (IHÉS). This was the brainchild of Léon Motchane, a French businessman of mixed Russian and Swiss parentage who believed that France needed a private, independent research establishment like the Institute for Advanced Study in Princeton. Grothendieck—30 years old at the time—was a founding professor at the IHÉS.

The privacy and independence of the IHÉS were eroded by its constant need for funding. Motchane's personal resources were not adequate, and from the mid-1960s he began to accept small grants from the French military. Grothendieck was a passionate anti-militarist. When he could not persuade Motchane to give up the military funding, he resigned from the institute in May 1970.

During those 12 years at the IHÉS, Grothendieck was a mathematical sensation. The field in which he worked was algebraic geometry, but he was able to raise the subject to such a level of generality as to take in key parts of number theory, topology, and analysis, too.

Here Grothendieck was following the pioneering work of a French mathematician of the previous generation, Jean Leray. Like Poncelet 130 years before him, Leray had worked out his most important ideas while a prisoner of war. His dates are 1906–1998, the same as André Weil's. As an officer in the French army, Leray was captured when the nation fell in 1940 and spent the whole of the rest of World War II in a camp near Allentsteig in northern Austria. Up to that point Leray's specialty had been hydrodynamics. In order to avoid having his expertise conscripted by the Germans for war work, however, he switched his interest to the most abstract field he knew, algebraic topology, and pushed homology theory into the new territory I described in §15.3. This was the territory in which, in the following generation, Grothendieck and his coeval Jean-Pierre Serre at the Collège de France made their names as explorers.

Grothendieck was a charismatic teacher, whose corps of devoted students in the 1960s was thought by some to resemble a cult. His mathematical style was not to everyone's taste, and disparaging comments about him are not hard to find. By the testimony of many first-class mathematicians who knew and worked with him, though, including some whose own styles are quite non-Grothendieckian, he was, in those glory years, a bubbling fount of mathematical creativity, throwing off startling insights, deep conjectures, and brilliant results nonstop. In 1966, he was awarded the Fields Medal, the highest and

most coveted prize for mathematical excellence. "Built on work of
Weil and Zariski and effected fundamental advances in algebraic geometry," goes the citation.

§15.7   I introduced Grothendieck with the archetypes Holy Fool and
Mad Genius. Though I myself cannot understand much of his work,
comments on that work by mathematicians such as Nick Katz,
Michael Artin, Barry Mazur, Pierre Deligne, Sir Michael Atiyah, and
Vladimir Voevodsky (there are three more Fields Medal winners in
that list) are sufficient to persuade me of his genius. What about the
holiness, folly, and madness?

The holiness and the folly are combined in a kind of childlike
innocence, which everyone who has known Grothendieck has remarked on. Not that there is anything childlike about the man physi-
cally. He is (or at any rate was, in his prime) large, handsome, and
strong and an excellent boxer. At a 1972 political demonstration in
Avignon, Grothendieck, who was 44 years old, knocked down two
police officers who tried to arrest him.

The intense concentration on his work, though—he seems, by all
accounts, to have thought about very little but mathematics all
through his 20s and 30s—left him unworldly and grossly ill-
informed. IHÉS professor Louis Michel recalls telling Grothendieck,
around 1970, that a certain conference was being sponsored by NATO.
Grothendieck looked puzzled. Did he know what NATO was? asked
Michel. "No."

His zone of ignorance extended into areas of mathematics he was
not interested in—which is to say all mathematics except for the most
utterly abstract reaches of algebra. He did not, for example, find numbers interesting. Mathematicians sometimes refer to the number 57—
the product of 3 and 19 and therefore not a prime number—as
"Grothendieck's prime." The story goes that Grothendieck was par-
ticipating in a mathematical discussion when one of the other par-

ticipants suggested they try out a procedure that had been suggested, applicable to all prime numbers, on some particular prime. "You mean an actual prime number?" Grothendieck asked. Yes, said the other, an actual prime number. "All right," said Grothendieck, "let's take 57."

Grothendieck's own miserable childhood undoubtedly contributed to his rather patchy understanding of the world and of mathematics. His parents were both eccentric rebels. His father, a Ukrainian Jew named Shapiro, born in 1889, spent all his life in the shadow world of anarchist politics, the world described in Victor Serge's memoirs. There were several spells in tsarist prisons, and Shapiro lost an arm in a suicide attempt while trying to escape from the tsar's police. Lenin's totalitarianism did not suit him any better than the tsar's authoritarianism had, so he left Russia for Germany in 1921, and made a living as a street photographer so as not to violate his anarchist principles by having an employer.

Alexander's mother, from whom the boy took his name, was Johanna Grothendieck, a Gentile girl from Hamburg in rebellion against her bourgeois upbringing, living in Berlin and doing occasional writing for left-wing newspapers. Alexander was born in that city, and his first language was German. By the outbreak of World War II the family was in Paris. Both parents, however, had fought on the Republican side in the Spanish civil war and so were regarded by the French wartime authorities as potential subversives. Grothendieck's father was interned; then, after France fell, he was shipped off to Auschwitz to be killed. Mother and son spent two years in internment camps; then Alexander was moved to a small town in southern France where the Resistance was strong. Life was precarious, and in his autobiography Grothendieck writes of periodic sweeps when he and other Jews would have to hide in the woods for several days at a time. He survived, though, managed to pick up some kind of education at the town schools, and at age 17 was reunited with his mother. Three years later, after some desultory courses at a provincial university, Grothendieck was advised by one of his instructors to go to Paris and study under the great algebraist Henri Cartan. Grothendieck did, and his mathematical career was under way.

Grothendieck's autobiography expresses reverence for both his parents. Certainly the man is, as his father was, absolutely unbending in his convictions. Though he accepted the Fields Medal in 1966, he refused to travel to the International Congress of Mathematicians to receive it, because the Congress was being held in Moscow that year, and Grothendieck objected to the militaristic policies of the Soviet Union. The following year he made a three-week trip to North Vietnam and lectured on category theory in the forests of that country, whither the Hanoi students had been evacuated to escape American bombing.

Eleven years later—he was then 60 years old—Grothendieck was awarded the Crafoord Prize by the Royal Swedish Academy of Sciences. This one he declined altogether. (It carried a $200,000 award—Grothendieck seems never to have paid the slightest attention to money.) In his explanatory letter, soon afterward reprinted in the French daily newspaper *Le Monde*, Grothendieck railed against the dismal ethical standards of mathematicians and scientists.

All of this was in the true anarchist spirit. None of it was informed by the sour anti-Americanism that was already beginning to be a feature of French intellectual life. Nor, the Vietnam trip notwithstanding, has Grothendieck been any particular fan of communism or the USSR, as a great many French intellectuals have, to their shame. As an anarchist and a political ignoramus, Grothendieck probably thinks that all political systems are equally wicked, all armies mere instruments of murder, all wealthy folk oppressors of the poor. Some of the postmodernist cant of the age seems to have seeped into his brain. In the autobiography he remarks that:

> [E]very science, when we understand it not as *an instrument of power and domination* but as an adventure in knowledge pursued by our species across the ages, is nothing but this harmony, more or

less vast, more or less rich from one epoch to another, which un-
furls over the course of generations and centuries, by the delicate
counterpoint of all the themes appearing in turn, as if summoned
from the void.

That is actually rather beautifully put, once you get past those itali-
cized words (whose italics are mine) from the po-mo phrasebook.

§15.8   Holiness and madness. After resigning from the IHÉS,
Grothendieck taught for two years at the Collège de France in Paris,
but he had a disconcerting habit of giving over his lectures to pacifist
rants. After a couple of attempts to start a commune—unsuccessful
for all the usual reasons[166]—in 1973, he took a position at the University of Montpellier, down on the Mediterranean coast, west of
Marseilles. This was, by French academic standards, an extraordinary
self-demotion. Most French academics spend years scheming to get a
position in Paris and then, having gotten one, would submit to tor-
ture rather than give it up. Grothendieck gave Paris up without a
blink. Nothing worldly ever seems to have meant much to him.

During his 15 years at Montpellier—he retired in 1988, at age
60—Grothendieck wrote his autobiography, *Reaping and Sowing*
(never published but widely circulated in manuscript), as well as
mathematical and philosophical books and articles. He learned to
drive, atrociously of course. He became a minor cult figure in Japan,
and parties of Buddhist monks came to visit him. He became "green,"
hooked into the environmentalist movement, and protested to the
authorities on this and numerous other topics.

In July 1990, two years after his retirement, Grothendieck asked a
friend to take custody of all his mathematical papers. Soon afterward,
early in 1991, he disappeared. Admirers eventually tracked him down
to a remote village in the Pyrenees, where he remains to this day.
Some sources say he has become a Buddhist, others that he spends his
time railing at the Devil and all his works. Roy Lisker, who visited
Grothendieck in his hermitage, reported in 2001 that:

Although direct communication with him is next to impossible, his neighbors in the village where he resides look after him. Thus, although he is known to come up with ideas like living on dandelion soup and nothing else, they see to it that he maintains a proper diet. These neighbors also maintain contact with . . . well-wishers in Paris and Montpellier, so one doesn't need to worry about him.[167]

The mathematics of Alexander Grothendieck's golden years remains, and those who understand it—of whom I cannot claim to be one—speak of it with awe and wonder.

§15.9   It is often said that a nation's written literary language is sometimes closer to, sometimes further from, the ordinary speech of the people. English was close in Chaucer's time, was more distant in the Augustan Age of the early 18th century, is closer again in our own era. In an analogous way, algebra has sometimes been closer to, sometimes further from, the practical world of science.

The very earliest algebra arose, as we have seen, from practical problems of measurement, timekeeping, and land surveying. (Though not really significant, there is a pleasing symmetry in my having been able quite naturally to mention land surveying in both the first and the last chapters of this book—see §15.3.) Diophantus and the medieval Muslim mathematicians added a layer of abstraction, departing sometimes from practical matters to deal with algebraic topics for their own intrinsic interest. This attitude was carried forward into the Renaissance and early-modern period, where purealgebraic inquiries into cubic and quartic equations generated great interest and eventually general solutions.

From the invention of modern literal symbolism in the decades around 1600 to the late 18th-century assault on the general quintic equation, the new symbolism was widely used to tackle practical problems in civil and military engineering, astronomy and navigation, accounting, and the rudimentary beginnings of statistics. Algebra was perhaps closer to the earthbound realm of practical affairs during this period than it had been since its origins in Mesopotamia.

The growth of pure algebra in the 19th century, however, was so abundant that the subject raced ahead of any practical applications to dwell almost alone in a realm of perfect uselessness. Even when practical folk took inspiration from algebra, they did so carelessly and uncomprehendingly. I have already mentioned (§8.7) the highly irreverent attitude of Gibbs and Heaviside to Hamilton's precious quaternions. By the end of the 19th century, algebra had left science far behind. The young David Hilbert would have laughed out loud if you had asked him, in 1893, to suggest some practical application of the Nullstellensatz.

The 20th century, for all its trend to yet higher abstraction, saw the gap close somewhat. All the new mathematical objects discovered in the 19th century have found some scientific application, if only in speculative theories. This is an aspect of the "miracle" that Eugene Wigner spoke of in his landmark 1960 essay, "The Unreasonable Effectiveness of Mathematics in the Natural Sciences." Somehow, these products of pure intellection, these groups and matrices, these fields and manifolds, turn out to be pictures of real things or real processes in the real world.

The "unreasonable effectiveness" of algebra has shown up all over the place. Groups, for example, are important in the theory of coding and encryption; matrices are now fundamental to economic analysis; notions from algebraic topology show up in areas from power generation to the design of computer chips. Even category theory has, according to its propagandists, worked wonders in the design of computer languages, though I cannot myself judge the value of this claim.

Undoubtedly, though, the most striking illustrations of Wigner's "unreasonable effectiveness," so far as algebra is concerned, have occurred in modern physics.

§15.10   The two great 20th-century revolutions in physics were of course those that go under the heading of relativity and quantum theory. Both depended on concepts from 19th-century "pure" algebra.

Item. In the special theory of relativity, measurements of time and space made in one frame of reference can be "translated" to measurements made in another (traveling, of course, at constant velocity relative to the first) by means of a Lorentz transformation. These transformations can be modeled as rotations of the coordinate system in a certain four-dimensional space—in other words, as a Lie group.

Item. In general relativity this four-dimensional space-time is distorted—curved—by the presence of matter and energy. For the proper description of it we must rely on the *tensor calculus*, developed by the Italian algebraic geometers out of the work begun by Hamilton, Riemann, and Grassmann.

Item. When the young physicist Werner Heisenberg, in the spring of 1925, was working on the radiation frequencies emitted by an atom that "jumps" from one quantum state to another, he found himself looking at large square arrays of numbers, the number in the nth column of the mth row in an array being the probability that the atom would jump from state m to state n. The logic of the situation required him to multiply these arrays together and suggested the only proper technique for doing so, but when he tried to carry out this multiplication, he found that it was noncommutative. Multiplying array A by array B gave one result; multiplying B by A gave a different result. What on earth was going on? Fortunately, Heisenberg was a research assistant at the University of Göttingen, so he had David Hilbert and Emmy Noether on hand to gently explain the principles of matrix algebra.

Item. By the early 1960s, physicists had uncovered a bewildering zoo of the type of nuclear particles called hadrons. Murray Gell-Mann, a young physicist at Caltech, noticed that the properties of the hadrons, though they did not follow any obvious linear pattern, made sense in the context of another Lie group, one that appears when we study rotations in a two-dimensional space whose coordinates are complex numbers. Working the data, Gell-Mann then saw that this original impression was superficial. The equivalent Lie group in a space of *three* complex dimensions had greater explanatory power. It required the existence of particles that had not yet been observed, though. Gell-Mann published his results, experimenters powered up their particle colliders, and the predicted particles were duly observed.[168]

Now, in the early 21st century, even stranger and bolder physical theories are circulating. None of them could have been conceived without the work of Hamilton and Grassmann, Cayley and Sylvester, Hilbert and Noether. The most adventurous of these theories arise from efforts to unify the two great 20th-century discoveries, relativity and quantum mechanics. They bear names such as string theory, supersymmetric string theory, M-theory, and loop quantum gravity. All draw at least some of their inspiration from 20th-century algebra or algebraic geometry.

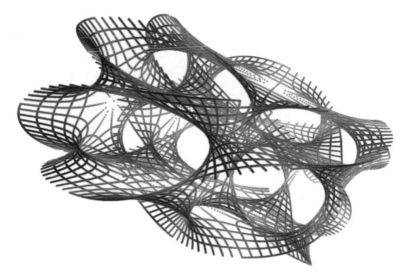

FIGURE 15-4 A Calabi–Yau manifold.

Take, for instance, the Calabi–Yau manifolds that provide the "missing" dimensions demanded by string theory. These are six-dimensional spaces that, according to string theory, lurk in the tiniest regions of space-time, down at the Planck length (that is, a billionth of a trillionth of a trillionth of a centimeter). They were first thought up by the German mathematician Erich Kähler (1902–2000), who, like Oscar Zariski, though a few years later (1932–1933), had studied in Rome with the Italian algebraic geometers.

Working from some ideas of Riemann's, Kähler defined a family of manifolds with certain general and interesting properties.[169] Every Riemann surface, for example, is a Kähler manifold. An American mathematician of the following generation, Eugenio Calabi identified a subclass of Kähler manifolds and conjectured that their curvature should have an interesting kind of simplicity.

Shing-tung Yau, a young mathematician from China, proved the Calabi conjecture in 1977, and these types of spaces are now called Calabi–Yau manifolds.[170] The simplicity of their curvature—a certain kind of "smoothness"—makes them ideal for the kinds of string motions that, according to string theory, appear to our instruments as all the many varieties of subatomic particles and forces, including gravitation. The fact of their being six-dimensional is a bit alarming, but these "extra" dimensions are "folded up" out of sight from our vantage point up here in the macroscopic world, just as a thick three-dimensional hawser looks one-dimensional when viewed from sufficiently far away.

§15.11   It seems, therefore, that there are reasons to think that the reaching up to ever higher levels of abstraction that characterized algebra in the 20th century may cease, or at least take a pause, while algebraists occupy themselves with answering puzzles posed by physicists, and while the proper status of hyper-abstract approaches such as category theory are sorted out.

It is also possible that algebra, as a separate discipline within mathematics, may not survive. The 20th century was a period of unification, with algebra invading other areas of math, and they counter-invading it. If I am engaged in the study of families of functions on multidimensional manifolds, those families having a group structure, am I working in analysis (the functions), topology (the manifolds), or algebra (the groups)?

The case for thinking that algebra will survive—a case I favor—rests on the idea that there is a distinctly algebraic way of thinking. We go back here (§14.3 again) to Hamilton's "Algebra as the Science of Pure Time" and to other speculations on the relationship between mathematical thinking and other kinds of mental activity. The great algebraist Sir Michael Atiyah, in a June 2000 lecture in Toronto, spoke of geometry and algebra as "the two formal pillars of mathematics" and argued that they belong to different regions of our minds.

> Geometry is . . . about space. . . . If I look out at the audience in this room I can see a lot, in one single second or microsecond I can take in a vast amount of information. . . . Algebra, on the other hand . . . is concerned essentially with time. Whatever kind of algebra you are doing, a sequence of operations is performed one after the other, and "one after the other" means you have got to have time. In a static universe you cannot imagine algebra, but geometry is essentially static.[171]

It is convenient to recall that the "sequence of operations" Sir Michael spoke of is known formally as an *algorithm* and that this word is (§3.5) a corrupted version of the name of the man who gave us that other word, *algebra.*

Sir Michael's train of thought was taken to its furthest extreme, so far as I know, by mathematician Eric Grunwald in the spring 2005 issue of *Mathematical Intelligencer.* Grunwald, under the essay heading "Evolution and Design Inside and Outside Mathematics," argues for a broad dichotomy in thinking, a sort of yin–yang binary scheme

I have sketched below. (Some of the entries are additions of my own, for which Grunwald should not be held responsible.)

Yin	*Yang*
geometry	algebra
discover	invent
sight	hearing
painting	music
prescriptive (lexicography)	descriptive
theory building	problem solving
security	adventure
patterns in space	processes through time
Newton	Leibniz[172]
Poincaré	Hilbert
Einstein[173]	Mach
design	evolution
socialism	capitalism
Platonic (view of mathematics)	"social construction"
theoretical (physics)	experimental

One can, of course, play these intellectual parlor games all night without coming to much in the way of conclusions. (I am mildly surprised at myself for having had sufficient power of self-control to leave "Augustinian" and "Pelagian" off the list.)

I do think, though, that Sir Michael and Grunwald are on to something. Mathematics today, at the highest levels, is wonderfully unified, with notions from one traditional field (geometry, number theory) flowing easily into another (algebra, analysis). There are still distinct styles of thinking, though, distinct ways of approaching problems and reaching new insights. We heard much talk a few years ago about whether the End of History had arrived. I can't recall whether our pundits and philosophers came to any conclusions about that larger matter, but I feel sure that algebra, at least, has not ended yet.

ENDNOTES

Introduction

1. Published in 2002 by BarCharts, Inc., of Boca Raton, FL. The author of both is credited as S. B. Kizlik.

2. "Invariants," *Duke Mathematical Journal*, 5:489–502.

3. I shall sometimes use "the 19th century," as historian John Lukacs does, to refer to the period from 1815 to 1914. Here, however, the ordinary calendrical sense is intended.

4. "Mathematical object" means a thing that is of professional interest to mathematicians, which they struggle to understand and develop theorems about. The mathematical objects most familiar to nonmathematicians are (1) numbers and (2) the points, lines, triangles, circles, cubes, etc., that dwell in the two- and three-dimensional spaces of Euclid's geometry.

5. Discovered or invented? My inclination is to take the "Platonic" view that these objects are in the world somewhere, waiting for human ingenuity to discover them. That is the frame of mind in which most mathematics is done by most professional mathematicians most of the time. The point is a nontrivial one, but it is only marginally relative to the history of algebra, so I shall say very little more about it.

Math Primer: Numbers and Polynomials

6. In modern usage, $\mathbb{N}$ most often includes the number zero. I am philosophically in sympathy with this. If you send me to peer into the next room, count the number of people in the room, and report the answer back to you, "zero" is a possible answer. Therefore zero ought to be included among the counting numbers. However, because my approach in this book is historical, I shall leave zero out of $\mathbb{N}$.

7. The common proof, first given by Euclid, argues *reductio ad absurdum*. Suppose the thing is *not* true. Suppose, that is, that some rational number $\frac{p}{q}$, with p and q both whole numbers, does indeed have the property that $\frac{p}{q} \times \frac{p}{q} = 2$. Assuming we have $\frac{p}{q}$ in its lowest terms (that is, with common factors canceled out top and bottom—a thing that can always be done), either p or q must be an odd number. Since multiplying both sides of the equation by q twice gives $p^2 = 2q^2$, and only even numbers have even squares, p must be even, so q must be odd. So p is $2k$, for some whole number k. But then $p^2 = 4k^2$, so $4k^2 = 2q^2$, so $q^2 = 2k^2$, and q must also be even. So q is odd and q is even—an absurdity. The premise, therefore, is false, and there is *no* rational number whose square is 2. (For a different proof, see Endnote 11 in my book *Prime Obsession*.)

8. Pythagoras's theorem concerns the lengths of the sides of a plane right-angled triangle. It is a matter of simple observation that the side *opposite* the right angle must be longer than either of the other two sides. The theorem asserts that the square of its length is equal to the squares of their lengths, added together: $c^2 = a^2 + b^2$, where a and b are the lengths of the sides forming the right angle and c the length of the side opposite it. Another way to say this, as in Figure NP-4, is $c = \sqrt{a^2 + b^2}$.

Chapter 1: Four Thousand Years Ago

9. The dating of early Mesopotamian history is still not settled. At the time of this writing, the "middle" chronology is the one most often cited, so that is the one I shall use. Also in play are the low, ultra-low, and high chronologies. An event placed at 2000 BCE in the middle chronology would be dated 2056 BCE in the high chronology, 1936

BCE in the low, and 1911 BCE in the ultra-low. No doubt friendships are shattered and marriages sundered by these disputes among professional Assyriologists. I have no strong opinion, and precise dates for this period are not important to my narrative. The much earlier dates found in materials written before about 1950 are at any rate now discredited.

10. The spelling "Hammurapi" is also common. Older English-language texts use "Khammurabi," "Ammurapi," and "Khammuram." The identification of Hammurabi with the Amraphel of Genesis 14:1 is, however, now out of favor. Abraham's dates are highly speculative, but no one seems to think he lived as late as Hammurabi's reign.

11. The *second* is more familiar to the Western tradition. It was by Nebuchadnezzar of the second Babylonian empire that the Jews were dragged off into captivity; Daniel served that same monarch; and the writing on the wall at Belshazzar's feast presaged the fall of the second Babylon to the Persians. All that was a thousand years later than the time of Hammurabi, though, and is not part of this story.

12. Key names here are the Dane Carsten Niebuhr, the German Georg Friedrich Grotefend, and the Englishman Sir Henry Rawlinson. Grotefend, by the way, was from the German state of Hanover and was engaged to the task of deciphering cuneiform by the great Hanoverian university of Göttingen, later famous as a center of mathematical excellence.

13. Cuneiform is not actually all that hard to read. The best short guide to cuneiform numeration is in John Conway and Richard Guy's *Book of Numbers.*

14. In case it is not familiar: The quadratic equation $x^2 + px + q = 0$ has two solutions, given by taking the $\pm$ ("plus or minus") sign to be either a plus or a minus in the formula

$$x = \frac{-p \pm \sqrt{p^2 - 4q}}{2}$$

Chapter 2: The Father of Algebra

15. Diophantus wrote his own name as "Diophantos," in the Greek form. His work became generally known to Europeans in a Latin translation, though, so the Latin form "Diophantus" has stuck.

16. So, for example, $\psi\mu\theta$ would be 749. The letters used for units can be recycled to show thousands: $\delta\psi\mu\theta$, for example, meaning 4,749. The δ, which normally means 4, is here being used to mean 4,000. To get beyond 9,999, digits were grouped in fours, separated by an M (for "myriad") or in Diophantus's notation by a dot. The number $\delta\tau o\beta \cdot \eta\,$⟩$\zeta$, for instance, would be 43,728,907. (That weird-looking letter ⟩ is one of the obsolete ones, a "san," here being used to mean 900. Since ⟩ stands for 7, ⟩ζ means 907. Note the absence of any positional zero, since with this method none is needed.)

17. The ς—a "terminal sigma"—had a little cross line at the top when used in this way. I haven't been able to duplicate this. The Michigan Papyrus dates from the early 2nd century CE, a century or so before the most popularly accepted dates for Diophantus.

Chapter 3: Completion and Reduction

18. Of Plotinus, the founder of this theory—another Alexandrian, by the way, and quite likely a contemporary of Diophantus—Bertrand Russell wrote: "Among the men who have been unhappy in a mundane sense, but resolutely determined to find a higher happiness in the world of theory, Plotinus holds a very high place." Neoplatonists thought very highly of mathematics, as of course did Plato and the original Platonists. Marinus, a later Neoplatonist, remarked: "I wish everything were mathematics."

19. The Jews must have returned, for the Muslim conqueror of the city in 640 CE reported that it contained "forty thousand tributary Jews."

20. "I am ignorant, and the assassins were probably regardless, whether their victim was yet alive," notes Gibbon (*The Decline and Fall of the Roman Empire*, Chapter 47). Charles Kingsley, of *Water Babies* fame, wrote a novel about Hypatia, in which she is still alive when the oyster

shells are applied. The novel is every bit as melodramatically Victorian as the Charles William Mitchell painting, which was inspired by it.

21. Here and in what follows I am using the word "Persian" in a loose way to refer to any of the peoples of present-day Iran and southern Central Asia who speak languages of the Indo-European family, excluding only the Armenians. There is really no satisfactory word here, "Aryan" having unpleasant connotations, "Iranian" belonging properly to a modern nation, not a subset of some language group. Plenty of these peoples would be unhappy to see themselves referred to as "Persians," and in a different historical context the term would cause confusion, but the poor writer must do his best.

22. Heraclius died 50 days later—"of a dropsy," says Gibbon.

23. "Amrou" in Gibbon.

24. Monophysites argued that the humanity and divinity of Christ are really just one thing. The opposite heresy—that they are two things—belonged to the Nestorians, who were so thoroughly banished from Christendom they ended up in China. You can see Nestorian crosses in the "forest of steles" in Xi'an city. The orthodox formula, adopted by the Council of Chalcedon in 451 and maintained by all the major Christian churches ever since, is that Christ's divinity and humanity are one thing and two things *at the same time*, "two natures without confusion, without change, without division, without separation."

25. The full name translates as "Father of Ja'far, Mohammed, son of Musa, the Khwarizmian." Khwarizm was an ancient state in what is now Uzbekistan. These Arabic names beginning with "al-" are usually indexed and cataloged under their second part, by the way. Al-Khwarizmi appears in the *DSB*, for example, among the Ks, not the As.

26. The Middle Ages began on Saturday, September 4, 476 CE, when the last emperor of Rome in the West was deposed by Odoacer the Barbarian. They ended on Tuesday, May 29, 1453, with the fall of Constantinople. The precise midpoint of the Middle Ages, if my numbers are right, was therefore at around midnight on Sunday, January 15, in the year 965.

27. These two factions of Shiites are sometimes referred to in English as "Twelvers" and "Seveners." Shiites believe that Ali, the fourth Caliph,

Mohammed's cousin and son-in-law, was the first Imam, a spiritual title of enormous authority. Several other Imams followed, the title being passed down from father to son. The line of succession was broken, though, and the split within the Shiites is over whether it was broken after Ismail, the seventh Imam, or Mahdi, the twelfth. The Ismailites are "Seveners." Most Shiites, nowadays practically all, are "Twelvers."

28. Malik Shah's names tell us something about the ethnic balance in the Seljuk empire. "Malik" and "Shah" are the Arabic and Persian words for "King." Malik Shah was actually a Turk, of course. Three ethnies in one person.

29. The Assassins were of the Ismailite confession and therefore at odds both with the Sunni rulers of the Seljuk empire and with other Shiites. Practicing a mystical approach to Islam that owed much to older Persian beliefs, they were persecuted by everybody and eventually retreated to a remote mountainous area of northern Iran, whence they carried out their horrible program of political murder. (The Crusaders, who knew them well, called Hasan Sabbah "the Old Man of the Mountains.") Murder aside, much of what has commonly been said about the Assassins is disputed by historians. There is, for example, no evidence that they fired up their killers with hashish, though they may have used the drug for religious purposes.

30. To cast the problem in modern terms: Suppose a sphere of diameter D, standing on a flat horizontal plane, is to have its top sliced off by a parallel horizontal plane at height x, in such a way that the remaining part of the sphere has R percent of the sphere's original volume. What should x be? The answer is found by solving the following cubic equation:

$$2\left(\frac{x}{D}\right)^3 - 3\left(\frac{x}{D}\right)^2 + \frac{R}{100} = 0$$

If R is 50 percent, for example, then of course x equal to half of D does the trick (that is, $x/D = 1/2$), since

$$2\left(\frac{1}{8}\right) - 3\left(\frac{1}{4}\right) + \frac{50}{100} = 0$$

31. A right-angled triangle with shorts sides 103 and 159 has, by Pythagoras's theorem (Endnote 8), a hypotenuse of length $\sqrt{103^2 + 159^2}$, which is $189.44656238655\ldots$. The length of the perpendicular is $86.44654088049\ldots$, so that if you add 103 to that, you do indeed get the hypotenuse, very nearly.

Math Primer: Cubic and Quartic Equations

32. Note that I assume the coefficient of x^3 is 1. There is no loss of generality in assuming this. A more general form would be $ax^3 + bx^2 + cx + d = 0$. Either a is zero, however, or it isn't. If it's zero, the equation isn't cubic; and if it's not, I can divide right through by it, reducing the coefficient of x^3 to 1.

33. You more often hear "reduced cubic" in our prosaic age. I prefer the older term.

34. An extremely confusing nomenclature, best restricted to this historical context. In the more general theory of equations, an irreducible equation is one that cannot be factored without enlarging your number field—going to a new "Russian doll." (See §FT.5 for more on this.) The cubic equation $x^3 - 7x + 6 = 0$ yields $q^3 + 4p^3/27$ equal to $-400/27$, so this is an "irreducible case." Yet $x^3 - 7x + 6$ factorizes very nicely, to $(x-1)(x-2)(x-3)$, so it is not irreducible in the proper sense. It is only that intermediate *quadratic* that is irreducible. Grrrr.

Chapter 4: Commerce and Competition

35. To prove this, call the nth term of the Fibonacci sequence u_n. So u_1 is 1, u_2 is also 1, u_3 is 2, u_4 is 3, and so on. Now construct this polynomial using the Fibonacci numbers as coefficients:

$$S = x^{n-1} + x^{n-2} + 2x^{n-3} + 3x^{n-4} + 5x^{n-5} + 8x^{n-6} + \cdots u_{n-2}x^2 + u_{n-1}x + u_n$$

Multiply both sides through by x to get xS. Repeat to get x^2S. Subtract S and xS from x^2S. You will see that, precisely because of the property of the Fibonacci sequence, most of the terms on the right-hand side disappear. The term in x^{n-6}, for example, will have coefficient $21 - 13 - 8$, which is zero. You are left with

$$(x^2 - x - 1)S = x^{n+1} - u_n x - u_{n-1} x - u_n$$

Setting x equal to each of the two roots of $x^2 - x - 1 = 0$ in turn eliminates S, giving you a pair of simultaneous equations in two unknowns, u_n and u_{n-1}. Eliminate u_{n-1} and the result follows.

36. The binomial theorem gives a formula for expanding $(a + b)^N$. In the particular case $N = 4$, it tells us that $(a + b)^4 = a^4 + 4a^3 b + 6a^2 b^2 + 4ab^3 + b^4$, and that's what I used here.

37. Not, as often written, *Liber abaci*, at any rate according to Kurt Vogel's *DSB* article on Fibonacci, to which I refer argumentative readers. The title translates as "The Book of Computation," not "The Book of the Abacus." As with the names of operas, the titles of books in Italian do not need to have every word capitalized.

38. He was born, in other words, within a year or two of the famous Leaning Tower, construction on which began in 1173, though it was not finished for 180 years. The lean became obvious almost at once, when the third story was reached.

39. Nowadays the town of Bejaïa (written "Bougie" in French) in Algeria, about 120 miles east of Algiers.

40. *Flos* is Latin for "flower," in the extended sense "the very best work of . . . ."

41. The scholar-statesman Michael Psellus, who served Byzantine emperors through the third quarter of the 11th century—he was prime minister under Michael VII (1071–1078)—certainly knew of Diophantus's literal symbolism.

42. Full title *Summa de arithmetica, geometria, proportioni et proportionalita*—"A Summary of Arithmetic, Geometry, Proportions, and Proportionalities." Note, by the way, that we have now passed into the era of printed books in Europe. Pacioli's, printed in Venice, was one of the first printed math books.

43. A later book of Pacioli's enjoyed the highly enviable distinction of having Leonardo da Vinci for its illustrator. Da Vinci and Pacioli were close friends. See Endnote 123 for another distinction of this sort. Yet an-

other of Pacioli's claims to fame is that of having coined the word "million."

44. The Italians had just spelled out "plus" and "minus" as *piu* and *meno*, respectively; though, like the powers of the unknown, these had increasingly been abbreviated, usually to "p." and "m."

45. The German algebraists of the 15th and 16th centuries were in fact called Cossists, and algebra "the Cossick art." The English mathematician Robert Recorde published a book in 1557 titled *The Whetstone of Witte, which is the second part of Arithmeticke, containing the Extraction of Roots, the Cossike Practice, with the Rules of Equation.* This was the first printed work to use the modern equals sign.

46. This book was not published in Cardano's lifetime. There is a translation of it included as an appendix to Oystein Ore's biography of Cardano mentioned below (Endnote 48).

47. Charles V is the ghostly monk in Verdi's opera *Don Carlos.* The title Holy Roman Emperor was elective, by the way. To secure it, Charles spent nearly a million ducats in bribes to the electors. He was the last emperor to be crowned by a pope (in Bologna, February 1530). Most of his contemporaries regarded him as king of Spain (the first such to have the name Charles and therefore sometimes confusingly referred to as Charles I of Spain), though he had been raised in Flanders and spoke Spanish poorly.

48. Oystein Ore, *Cardano, the Gambling Scholar* (1953). Ore's is, by the way, the most readable book-length account of Cardano I have seen, though unfortunately long out of print. For a very detailed account of Cardano's astrology, see Anthony Grafton's *Cardano's Cosmos* (1999). There are numerous other books about Cardano, including at least three other biographies.

49. They are given in detail in Ore's book. Ore gives over 55 pages to the Cardano–Tartaglia affair, which I have condensed here into a few paragraphs. It is well worth reading in full. For another full account, though with facts and dates varying slightly from Ore's (whose I have used here), see Martin A. Nordgaard's "Sidelights on the Cardan-Tartaglia Controversy, in *National Mathematics Magazine* 13 (1937–1938): 327–346, reprinted by the Mathematical Association of America in their

2004 book *Sherlock Holmes in Babylon*, M. Anderson, V. Katz, and R. Wilson, Eds.

Chapter 5: Relief for the Imagination

50. English-speaking mathematicians generally pronounce Viète's name "Vee-et," the more stubbornly anglophone tending toward the name of a well-known vegetable-juice drink. The name is sometimes written in the Latinized form "Vieta."

51. Commonly but not altogether accurately. The Huguenots were Calvinists. Not all French Protestants were, and there must have been many who would not have thanked you for referring to them as Huguenots. The name has stuck, though, and for passing reference in a book of this kind, "Huguenot" can be taken as a synonym for "early French Protestant." The etymology of the word is obscure.

52. Though the English, for once, were on good diplomatic terms with France all through this period, the blizzard of anti-French jokes and insults in Shakespeare's *Henry VI Part I* (1592) notwithstanding.

53. "All of Viète's mathematical investigations are clearly connected to his astronomical and cosmological work"—H. L. L. Busard in the *DSB*. The astronomy-trigonometry connection comes from dealing with the celestial sphere, computing and predicting the altitudes of stars, and so on.

54. Which was very horrible. Harriot got a cancer in his nose, perhaps from the new habit of smoking tobacco he had picked up in Virginia, and his face was gradually eaten away over the last eight years of his life.

Chapter 6: The Lion's Claw

55. In the old calendar, which was scrapped in 1752. According to the calendar we currently use, his birth date was January 4, 1643. This is why the date is sometimes given as the one year, sometimes as the other.

56. A review of Patricia Fara's book *Newton: The Making of Genius*, in *The New Criterion*, May 2003. In saying that Newton had "no interest in public affairs," I was referring to the momentous political events of his

time. He was master of the Royal Mint from 1696 onward, carrying out his duties diligently and imaginatively. He was also active in the Royal Society, of which he was elected and reelected president every year from 1703 until his death. *And,* he stood up courageously for his university against the sectarian bullying of King James II. For all that, I seriously doubt Newton ever spent five minutes together thinking about politics, about national or international affairs.

57. Or "Sir Isaac," if you like. He was knighted by Queen Anne in 1705, the first scientist ever to be so honored. In all strictness he should be referred to as "Newton" for his deeds before that date, "Sir Isaac" afterward. Nobody can be bothered to be so punctilious, though, and I am not going to be the one to set a precedent.

58. Neither the Latin original nor the English translation can easily be found. The text of the book, however, is included in volume 2 of *The Mathematical Papers of Isaac Newton* (D. T. Whiteside, ed., 1967).

59. I especially recommend Michael Artin's textbook *Algebra*, pp. 527–530 in my edition (1991), which does the whole thing as clearly as it can be done.

60. As Gauss, and later Kronecker, pointed out, there are some deep philosophical issues involved here. For a thorough discussion, see Harold Edwards's book *Galois Theory*, §§49–61.

Math Primer: Roots of Unity

61. The word "cyclotomic" seems to have been first used in this context by J. J. Sylvester in 1879.

62. "Primitive nth root of unity" should not be confused with the number-theory term of art "primitive root of a prime number." A number g is a primitive root of a prime number p if $g, g^2, g^3, g^4, \ldots, g^{p-1}$, when you take their remainders after division by p, are $1, 2, 3, \ldots, p-1$, in some order. For example, 8 is a primitive root of 11. If you take the powers of 8, from the first to the 10th, you get 8, 64, 512, 4096, 32768, 262144, 2097152, 16777216, 134217728, and 1073741824. Taking remainders after division by 11: 8, 9, 6, 4, 10, 3, 2, 5, 7, and 1. So 8 is a primitive root of 11. On the other hand, 3 is *not* a primitive root of 11. The first 10

powers of 3 are 3, 9, 27, 81, 243, 729, 2187, 6561, 19683, and 59049. Dividing by 11 and taking remainders: 3, 9, 5, 4, 1, 3, 9, 5, 4, and 1. Not a primitive root. This concept of primitive root is in fact *related* to the one in my main text, but it is not the same. Since 11 is a prime number, every 11th root of unity is a primitive 11th root of unity, but the primitive roots of 11 in a number-theoretic sense are only 2, 6, 7, and 8.

Incidentally, I can now explain that "more restricted sense" of the term "cyclotomic equation." It is the equation whose solutions are all the *primitive n*th roots of unity. So in the case $n = 6$, it would be the equation $(x + \omega)(x + \omega^2) = 0$, that being the equation with solutions $x = -\omega$ and $x = -\omega^2$. This equation multiplies out as $x^2 - x + 1 = 0$.

Chapter 7: The Assault on the Quintic

63. William Dunham's book *Euler, The Master of Us All* (1999) manages to do justice to both the man and his mathematics.

64. More properly, the Académie des Sciences, founded in Paris in 1666 by Jean-Baptiste Colbert, part of the great awakening of European science in the late 17th century. Compare Britain's Royal Society, 1660. The Académie used to meet in the Louvre.

65. *Galois Theory*, p. 19.

66. Lagrange is one of the greats, with index score 30 in Charles Murray's scoring (*Human Accomplishment*, 2003). Euler leads the field with an index score of 100. Newton has 89, Euclid 83, Gauss 81, Cauchy 34. Poor Vandermonde has index score only 1, and that is probably on account of "his" determinant.

67. I am simplifying here to the point of falsehood. Instead of "polynomial," I should really say "rational function." I'm going to explain that when I get to field theory, though. "Polynomial" will do for the time being.

68. Either J. J. O'Connor or E. F. Robertson, joint authors of the article on Ruffini at the indispensable math Web site of the University of St. Andrews in Scotland, *www-groups.dcs.st-andrews.ac.uk/~history/index.html.*

69. Ruffini was a licensed medical practitioner as well as a mathematician. This fact gives me an excuse to mention another 18th-century algebraist with this same dual qualification, though from the generation before Ruffini. This was the Englishman Edward Waring (1736–1798). Waring took up Isaac Newton's chair as Lucasian Professor of Mathematics at Cambridge University in 1760. Seven years later, while still holding the chair, he graduated with an M.D. degree He seems not to have practiced medicine much, though. His 1762 book *Miscellanea analytica* gave a treatment of the relations between symmetric functions of an equation's solutions, and the equation's coefficients—the topic I covered in relation to Isaac Newton's jottings. (The second edition of Waring's book is confusingly called *Meditationes algebraicae.*) While I am filling in like this, I may as well note the achievement of Swedish mathematician Erland Bring, who in 1786 figured out that any quintic equation can be reduced to one with no second, third, or fourth power of the unknown, in other words to one like this: $x^5 + px + q = 0$. I should like to call this a "severely depressed quintic."

70. It seems that Cauchy actually believed in the medieval theory of the Divine Right of Kings, often mistakenly thought of as a Protestant doctrine but in fact going back to medieval times and popular in 17th-century France. If this is right, Cauchy must have been the last person of any intellectual eminence to adhere to this theory.

71. I have taken this from Peter Pesic's fine short book *Abel's Proof* (2003). E. T. Bell, however, gives the number of Abel children as seven. Bell's chapter on Abel in *Men of Mathematics*, by the way, is worth reading just as a piece of 1930s Americana. It is Bell at his best—or, depending on your tolerance for writers chewing the scenery, his worst.

72. The spelling was later changed to conform to a more authentically Norwegian orthography: Kristiania.

73. You don't even have to be a mathematician. After the publication of my book about the Riemann hypothesis, I got a steady trickle of letters and e-mails from people claiming to have resolved that very profound mystery. Wishing neither to scrutinize their work nor to appear unkind, I developed a stock response along the following lines: "I am not a working mathematician, only a writer with a math degree. The fact of my having written a book about the Riemann hypothesis does not

qualify me to pass judgment on work in this area. I once wrote a book about opera, but I cannot sing. I suggest you get in touch with the math department at your local university."

74. A detailed proof takes us deeper than I want to go in this text. I refer curious readers to Peter Pesic's book *Abel's Proof*, which does the thing in as elementary a way as it can be done, I think, and at three different levels: an overview (more detailed than mine), Abel's actual 1824 paper, and some explanation of missing logic steps in the paper. Van der Waerden's *History of Algebra* also gives a neat 2½-page summary, though at a higher level.

Chapter 8: The Leap into the Fourth Dimension

75. A great favorite of mine, Dewdney's book is a wonder and well worth reading as an imaginative exercise. How, for example, does a two-dimensional creature lock his door? And if he has an alimentary canal running through him from one end to the other, what prevents him from falling into two separate pieces?

76. This story can be found in *Mathenauts*, a 1987 anthology of math-related science fiction stories edited by Rucker himself. Of the 23 stories in this collection, more than half make some play on the idea of a fourth dimension—about average for mathematical science fiction, in my experience.

77. No conscientious novelist of the 1990s thought his book was complete unless it included a reference to Heisenberg's uncertainty principle, which Heisenberg first stated in 1927.

78. There is a large literature on Hamilton, including at least three full-scale biographies. I have depended mainly on the 1980 biography by Thomas Hankins, supplemented by some references in mathematical magazines, textbooks, and Web sites.

79. Claims like this seem to have been common in early 19th-century England and America. The writer George Borrow (*The Bible in Spain*, *Lavengro*), born two years before Hamilton, is likewise supposed to have been the master of numerous languages—Dr. Ann Ridler, who has made a study of Borrow's linguistic skills, lists him as having pos-

sessed reading competence in 51 languages and dialects. Dr. Ridler also, in this context, mentions the American writer John Neal, born 1793, author of *Brother Jonathan*, who claimed that: "In the course of two or three years [I] made myself pretty well acquainted with French, Spanish, Italian, Portuguese, German, Swedish, Danish, beside overhauling the Hebrew, Latin, Greek, and Saxon . . ." The poet Longfellow, born a year and a half after Hamilton, was appointed professor of modern languages at Bowdoin when only 19 years old, provided he actually master some modern languages. He promptly taught himself French, Spanish, Italian, and German to a good degree of reading competency—we have independent confirmations of this—in 9, 9, 12, and 6 months, respectively, between 1826 and 1829. Writing as a person who, in spite of valiant personal struggles and the dogged efforts of several excellent teachers, has failed to master even one foreign language, I am baffled by all this. Perhaps there was something in the water back then.

80. It is natural to wonder whether Catherine was related to Walt Disney. Perhaps she was, but I have not been able to find any connection. It is an old family (originally Norman French "D'Isigney") with a large Irish branch. Walt descended from Arundel Elias Disney, born in Ireland about 1803. The story that Walt was an illegitimate child of Spanish parentage, adopted into the Disneys, is an urban legend.

81. In what is now the drab Dublin Industrial Estate, about three miles northwest of the city center.

82. We now know that Gauss had conceived of a noncommutative algebra as far back as 1820 but had not bothered to publish his thoughts. You had to get up very early in the morning to be up before Gauss.

83. Octonions were independently discovered by Cayley in 1845, and are sometimes called Cayley numbers.

84. Well, is it? Not in any simply geometrical sense. There is no "fourth direction" in which, by a supreme effort of will and imagination, you might move yourself, thereby leaving our three-dimensional world. If you did so, you would be destroyed at once, because even the simplest physical laws—the inverse square law, for instance—lead to very unpleasant consequences if you try to embed them in a four-dimensional Euclidean geometry. It is of course true that the space–time of modern

physics is conveniently described by a four-dimensional geometry, but that geometry is radically non-Euclidean, so you should banish from your mind any thought of taking an ordinary Euclidean trip through it. The human imagination is a very mysterious thing, though. The late H. S. M. Coxeter, in his book *Regular Polytopes*, notes of his friend John Flinders Petrie that: "In periods of intense concentration he could answer questions about complicated four-dimensional figures by 'visualizing' them."

85.   There is a counter-argument to be made in respect of continental Europe, where "lines of force" arguments of the kind favored by Faraday were less popular than the older "action at a distance" ideas. Still, reading the mathematics of the time, including the German mathematics, you can see that ideas about directional flows on surfaces and in space are just below the surface of the writers' minds.

86.   Quaternions have some minor application in quantum theory. I quote from some notes passed on to me by a helpful physicist friend: "Interestingly, if one formulates the rotation kinematics in terms of quaternions, the resultant 7×7 covariance matrix (the solution of the Riccati equation) is singular, because of the linear dependence of the 4-parameter Euler symmetric parameters." Just so. Conway and Smith's 2002 book, *On Quaternions and Octonions*, offers a very comprehensive coverage of Hamilton's brainchild, but the math is at a high level. Professor Andrew J. Hanson of Indiana University has a book titled *Visualizing Quaternions* coming out at about the same time as *Unknown Quantity*, early in 2006. I have not seen this book but it promises a full account of, among many other things, the application of quaternions to computer animation.

Chapter 9: An Oblong Arrangement of Terms

87.   The regions of the modern People's Republic not included under Han rule were the southern and southeastern strip of provinces from Fujian to Yunnan; the two outer provinces of Manchuria; and all the western and northwestern territories acquired during the modern period, with non-Chinese (Turkic, Tibetan, Mongolian) base populations.

88. That is one reason the ancient Chinese written language has such a severely abbreviated style. The classical texts were not so much narrative as mnemonic. *Shen zhong zhui yuan*, Confucius tells us (*Analects*, 1.ix). James Legge translates this as: "Let there be a careful attention to perform the funeral rites to parents, and let them be followed when long gone with the ceremonies of sacrifice." That's four Chinese syllables to 39 English ones.

89. This calendar was the work of Luoxia Hong, who lived about 130–70 BCE.

90. George MacDonald Ross, *Leibniz* (Oxford University Press, 1984).

91. Bernoulli numbers turn up when you try to get formulas for the sums of whole-number powers, like $1^5 + 2^5 + 3^5 + \ldots + n^5$. The precise way they turn up would take too long to explain here; there is a good discussion in Conway and Guy's *Book of Numbers*. The first few Bernoulli numbers, starting with B_0, are: $1, -\frac{1}{2}, \frac{1}{6}, 0, -\frac{1}{30}, 0, \frac{1}{42}, 0, -\frac{1}{30}$ (yes, again), $0, \frac{5}{66}, 0, -\frac{691}{2730}, 0, \frac{7}{6}, 0, -\frac{3617}{510}, 0, \frac{43867}{798}, \ldots$. Notice that all the odd-numbered Bernoulli numbers after B_1 are zero. Bernoulli numbers make another brief appearance in Chapter 12.

92. More observations give you better accuracy. Furthermore, the planets are perturbed out of their ideal second-degree curves by each other's gravitational influence. This accounts for Gauss using six observations on Pallas. Did Gauss know Cramer's rule? Certainly, but for these ad hoc calculations, the less general elimination method was perfectly adequate.

93. As an undergraduate I was taught to think of this as "diving rows into columns." To calculate the element located where the *m*th row of the product matrix meets the *n*th column, you take the *m*th row of the first matrix, "tip" it through 90 degrees clockwise, then drop it down alongside the *n*th column of the second matrix. Multiplying matched-off pairs of numbers and adding up the products gives you the element. Here, for example, is a matrix product, written with proper matrix notation:

$$\begin{pmatrix} 1 & 2 & -1 \\ 3 & 8 & 2 \\ 4 & 9 & -1 \end{pmatrix} \times \begin{pmatrix} 1 & -1 & 4 \\ 7 & 6 & -2 \\ 5 & -1 & -5 \end{pmatrix} = \begin{pmatrix} 10 & 12 & 5 \\ 69 & 43 & -14 \\ 62 & 51 & 3 \end{pmatrix}$$

To get that −14 in the second row, third column of the answer, I took the second row from the first matrix (3, 8, 2), then the third column from the second matrix (4, −2, −5), then "dived" the row into the column to calculate $3 \times 4 + 8 \times (−2) + 2 \times (−5) = −14$. That's all there is to matrix multiplication. You might want to try multiplying the two matrices in the other order to confirm that matrix multiplication is not, in general, commutative. To start you off: diving the first row of the *second* matrix into the first column of the *first*: $1 \times 1 + (−1) \times 3 + 4 \times 4 = 14$, so the top left number in the product matrix will be 14, not 10.

94. A matrix need not be square. If you think about that rule of multiplication—a row from the first matrix combining with a column from the second—you can see that as long as the number of *columns* in the first matrix is equal to the number of *rows* in the second, the multiplication can work. In fact, a matrix with m rows and n columns can multiply a matrix with n rows and p columns; the product will be a matrix with m rows and p columns. A very common case has $p = 1$. Any decent undergraduate textbook of modern algebra will clarify the issue. As always, I recommend Michael Artin's *Algebra*. Frank Ayres, Jr.'s book *Matrices*, in the Schaum's Outline Series, is also very good.

Chapter 10: Victoria's Brumous Isles

95. See Endnote 56.

96. Commenting on a different drinking song on a similar theme in his *Budget of Paradoxes*, De Morgan notes that "in 1800 a compliment to Newton without a fling at Descartes would have been held a lopsided structure."

97. And British affection for it lingered, at least in school textbooks. At a good British boys' school in the early 1960s, I learned physics and applied mathematics with the Newtonian dot notation.

98. The Scot I quoted, Duncan Gregory, only committed himself to mathematics at about the time of that remark and died less than four years later. He was a major influence on Boole, though. In fact, I lifted that Duncan quote not from its original publication (*Transactions of the*

Royal Society of Edinburgh, 14: 208–216) but from a paper ("On a General Method in Analysis") that Boole presented to the Royal Society in 1844.

99. Not to mention the income tax, which had been brought in as an emergency revenue source during the wars against Napoleon. The wars being over, the reformer Henry Brougham persuaded Parliament that the tax was no longer necessary, and it was duly abolished in 1816, to the horror of the government but to general rejoicing among the people.

100. "This place [the University of London] was founded by Jews and Welshmen," I was told when I first showed up on its doorstep. In fact, James Mill, Thomas Campbell, and Henry Brougham, the moving spirits behind the university's founding, were all Scottish. Financing for the project, however, was raised from the merchant classes of the city, who were indeed largely Methodists and Jews.

101. Two closely related functions, in fact. They are solutions of the ordinary differential equation

$$\frac{d^2y}{dx^2} = xy$$

and show up in several branches of physics.

102. You need to be born in a year numbered $N^2 - N$ to share this distinction. De Morgan was born in 1806 ($N = 43$). Subsequent lucky birth years are 1892, 1980, and 2070.

103. By Jevons, for example. See the article on De Morgan in the 1911 *Britannica*.

104. Which is to say the year of onset of the great and terrible potato famine. I do not believe the words "sensitive" or "intelligent" have ever been truthfully applied to any British government policy on Ireland. In establishing these new colleges, however, it must be said that the British were at least trying. In Ireland, even more than England, there was a clamor for nondenominational universities, open to anyone. The new colleges were a response to that clamor.

105. During the 1930s—when she was in her 70s—Alicia worked with the great geometer H. S. M. Coxeter (1907–2002). Coxeter has a long note on her in his book *Regular Polytopes*: "Her father . . . died when she was four years old, so her mathematical ability was purely hereditary. . . .

There was no possibility of education in the ordinary sense, but Mrs. Boole's friendship with [mystic, physician, eccentric, and social radical] James Hinton attracted to the house a continual stream of social crusaders and cranks. It was during those years that Hinton's son Howard brought a lot of small wooden cubes, and set the youngest three [Boole] girls the task of memorizing the arbitrary list of Latin words by which he named them, and piling them into shapes. [This] inspired Alice [*sic*] (at the age of about eighteen) to an extraordinarily intimate grasp of four-dimensional geometry . . . ." That Howard, by the way—full name Charles Howard Hinton—was the author of some speculations on the fourth dimension that may have helped inspire Abbott's *Flatland*.

Math Primer: Field Theory

106.   Not that there aren't deep and difficult results in field theory, but they don't lend themselves to directly algebraic methods so easily as group problems do and are usually tackled via algebraic geometry. Unfortunately, the term "field theory" has two utterly different meanings in math. It may mean what it means here: the study of that algebraic object called a "field." Or it may refer to the study of spaces at each point of which some quantity—a scalar, a vector, or something even more exotic—is defined. If I say "electromagnetic field theory," you will see what I mean.

107.   In fact, some textbook authors—Michael Artin is an example—prefer to write the elements of this field not as 0, 1, and 2 but instead as 0, 1, and -1. The arithmetic is then not *quite* so counterintuitive: Instead of $1 + 2 = 0$ you have $1 + (-1) = 0$. You are still stuck with $(-1) + (-1) = 1$, though.

108.   It is also one much improved with hindsight—a modern treatment, in fact. Galois's original 1830 memoir—it is reproduced as an appendix in Professor Edwards's book *Galois Theory*—does not employ the word "field." The word did not gain its algebraic sense until 1879, when Richard Dedekind first used it. My example, by the way, shows that F_9 can be constructed by appending the solution of an irreducible

quadratic equation to F_3 This is a particular case of a general theorem: If $q = p^n$, F_q can be constructed by appending to F_p some solution of an irreducible equation of the nth degree.

Chapter 11: Pistols at Dawn

109. The Web site is *dilip.chem.wfu.edu/Rothman/galois.html*

110. *Journal de Mathématiques Pures et Appliquées*, though called *Journal de Liouville* in its early years. Founded in 1836 and still going strong, it boasts itself "the second oldest mathematical journal in the world"— the oldest being Crelle's, started in 1826.

111. Somehow I have forgotten to mention that commutative groups are now called *Abelian*, in honor of a theorem of Abel's. Hence the hoary old mathematical joke: "Q—What is purple and commutes? A—An Abelian grape." It is also customary, when dealing with Abelian groups, to represent the group operation by addition, instead of by the more usual multiplication. The identity element for an Abelian group is therefore often represented by 0 (because $0 + a = a$ for all a), and the inverse of an element a is written as $-a$. I shall ignore all this in what follows, to keep things simple.

112. Richard Dedekind gave some lectures on Galois theory at Göttingen in the later 1850s.

113. More precisely, D_3 and S_3 are both instances of the same abstract group. The one and only abstract group of order 2, illustrated by my Figure FT-3, has not only D_2 and S_2 as instances but also C_2. Strictly speaking, all such notations as D_3, S_3, and C_2 name particular instances of abstract groups, and we should eschew phrases like "the group S_3," in favor of "the group of which S_3 is the most familiar instance," but no-one can be bothered to speak that strictly.

114. I shall not cover it in any more detail. For a very full and lucid account, see Keith Devlin's 1999 book, *Mathematics, The New Golden Age*. For a look at the final tally, though presented at a high level, see *The Atlas of Finite Groups* by J. H. Conway *et al.*, published by the Clarendon Press, Oxford (1985).

Chapter 12: Lady of the Rings

115. This resemblance between integers and polynomials was first noticed, or at any rate first remarked on, by the Dutch algebraist Simon Stevin around 1585. Stevin, by the way—I am sorry I have not found room for him in my main text—was a great propagandist for decimals and did much to make them known in Europe. His book on the subject inspired Thomas Jefferson to propose a decimal currency for the newborn United States, and it is to him (indirectly) that we owe the word "dime."

116. Howls of outrage from professional algebraists here. Yes, I am oversimplifying, though only by a little. In fact, the algebraic notion of a ring is somewhat broader than is implied by the examples I have given. A ring need not, for instance, have a multiplicative identity—that is, a "one"— which both $\mathbb{Z}$ and the polynomial ring have. And while addition must be commutative, multiplication need not be. This is not a textbook, though; I just want to get the general idea across.

117. As an example of the counterintuitive surprises that ring theory throws up, note that in the ring of numbers having the form $a + b\sqrt{5}$, where a and b are ordinary integers, the number $9 + 4\sqrt{5}$ is a unit. It divides into 1 exactly in this ring. Try it.

118. There is no easy way to define regular primes. The least difficult way is as follows. A prime p is regular if it divides exactly into *none* of the numerators of the Bernoulli numbers $B_{10}, B_{12}, B_{14}, B_{16}, \ldots, B_{p-3}$. (I have notes on the Bernoulli numbers in §9.3 and Endnote 91.) For example: Is 19 a regular prime? Only if it does not divide into any of the numerators of the numbers B_{10}, B_{12}, B_{14}, and B_{16}. Those numerators are 5, 691, 7, and 3617, and 19 indeed does not divide into any of them. Therefore 19 is a regular prime. The first irregular prime is 37, which divides exactly into the numerator of B_{32}, that numerator being 7,709,321,041,217.

119. At the time of this writing (April 2005), there has just been an anti-Japanese riot in Beijing, over similar indignities inflicted on China by Japan *60* years ago.

120. The town is now in western Poland and renamed Zary. Similarly, Breslau is now the city of Wroclaw in Poland. The entire German-Polish border was shifted westward after World War II.

121. With a dusting of good old Teutonic romanticism. "Poets are we" (Kronecker). "A mathematician who does not at the same time have some of the poet in him, will never be a mathematician" (Weierstrass)—*und so weiter.*

122. *Proof.* Suppose this is not so. Suppose there is some integer k that is not equal to $15m + 22n$ for any integers m and n whatsoever. Rewrite $15m + 22n$ as $15m + (15 + 7)n$, which is to say as $15(m + n) + 7n$. Then k can't be represented that way either—as 15 of something plus 7 of something. But look what I did: I replaced the original pair (15 and 22) with a new pair (the *lesser* of the original pair and the *difference* of the original pair: 15 and 7). Plainly I can keep doing that in a "method of descent" until I bump up against something solid. It is a matter of elementary arithmetic, proved by Euclid, that if I do so, the pair I shall eventually arrive at is the pair $(d, 0)$, where d is the greatest common divisor of my original two numbers. The g.c.d. of 15 and 22 is 1, so my argument ends up by asserting that k cannot be equal to $1 \times m + 0 \times n$, for any m and n whatsoever. That is nonsense, of course: $k = 1 \times k + 0 \times 0$. The result follows from *reductio ad absurdum.*

123. I have depended on the biography by J. Hannak, *Emanuel Lasker: The Life of a Chess Master* (1959), which I have been told is definitive. My copy—it is Heinrich Fraenkel's 1959 translation—includes a foreword by Albert Einstein. That is almost as enviable as having Leonardo da Vinci as your book's illustrator (see Endnote 43).

124. There is a Penguin Classics translation by Douglas Parmée. Rainer Werner Fassbinder made a very atmospheric movie version in 1974, with Hanna Schygulla as Effi and Wolfgang Schenck as her husband, Baron von Instetten.

125. *Aber meine Herren, wir sind doch in einer Universität und nicht in einer Badeanstalt.* You can't help but like Hilbert. The standard English-language biography of him is by Constance Reid (1970).

126. To a modern algebraist, "commutative" and "noncommutative" name two different *flavors* of algebra, leading to different kinds of applications. I can't hope to transmit this difference of flavor in an outline history of this kind, so I am not going to dwell on the commutative/noncommutative split any more than necessary to get across basic concepts.

127. Though for reasons I do not know it was published as a letter to the editor: "The Late Emmy Noether," *New York Times*, May 5, 1935.

Math Primer: Algebraic Geometry

128. The words "ellipse," "parabola," and "hyperbola" were all given to us by Appollonius.

129. Points at infinity were actually introduced into math by the astronomer Johannes Kepler around 1610, though the concept must have occurred to the painters of the Renaissance when they solved the problem of perspective. Kepler conceived of a straight line as being a circle whose center happened to be at infinity, a notion he probably acquired from his work with optics, where it occurs quite naturally.

130. Actually, not all authors follow this usage. Miles Reid, for example, in his otherwise excellent book *Undergraduate Algebraic Geometry* (London Mathematical Society Student Texts #12, Cambridge University Press, 1988), writes the general inhomogeneous quadratic polynomial as $ax^2 + bxy + cy^2 + dx + ey + f$.

131. The University of Minnesota's Geometry Center sells a video, *Outside In*, demonstrating one of the 20th century's most fascinating discoveries in topology: how to turn a sphere inside out. There is a brief animation on the Internet, but if you want to learn a little topology, I recommend buying the entire video. For a time I used to bring it out and play it to dinner guests as a conversation piece, but this was not an unqualified social success.

132. What it gets you is a *Riemann sphere*, a useful aid in thinking about functions of a complex variable, which acknowledge only one point at infinity.

Chapter 13: Geometry Makes a Comeback

133. Kant's ideas were the ultimate source of the analytic/synthetic dichotomy in geometry. Kant distinguished between analytic facts, whose truth can be demonstrated by pure logic, without any reference to the outside world, and synthetic facts, which are known by some other

means. Up to Kant, philosophers had assumed that the "other means" meant actual experience of our interacting with the world. Kant, however, denied this. In his metaphysics, there are truths that are not analytic yet are independent of experience. He thought that the facts of Euclidean geometry were of this kind—synthetic yet not derived from experience. Hence the connection between classical Greek math and the "synthetic" geometry of the early 19th century, though I have omitted several intermediate steps in the connection.

134. Oh dear. Cissoids, conchoids, epitrochoids, limaçons, and lemniscates are all particular types of curves. A lemniscate, for instance, is a figure-eight shape. Cusps are pointed bits of curves—the number 3, as usually written, has a cusp in its middle. Nodes are places where a curve crosses itself; there is one in the middle of a lemniscate. Cayleyans, Hessians, and Steinerians are curves that can be derived from a given curve by various maneuvers.

135. There are in fact a number of ways to "realize" homogeneous coordinates for two-dimensional geometry. One realization is areal coordinates (pronounced "AH-ree-ul"). Pick three lines in the plane forming a triangle. From any point, draw straight lines to the three corners of the triangle. This gives three new triangles, each having your chosen point at one vertex, opposite a side of the base triangle. The three areas of these triangles, appropriately signed, work very well as a system of homogeneous coordinates. Areal coordinates are a tidied-up version of Möbius's barycentric coordinates, in which a point is defined by the three weights that would need to be placed at the vertices of a base triangle in order for the chosen point to be their center of mass. Similar arrangements can be made in spaces of more than two dimensions, though of course the algebra gets more complicated *really* fast.

136. The 1888 result is properly called Hilbert's Basis Theorem and can be found under that name in any good textbook of higher algebra or modern algebraic geometry.

137. For mathematically well-equipped students, I recommend *An Invitation to Algebraic Geometry*, by Smith, Kahanpää, Kekäläinen, and Traves (Springer, 2000). This book covers all the essentials in an up-to-date style and has *plenty of exercises*! The Nullstellensatz is on page 21.

138. "I don't know the origin of this unattractive term," says Michael Artin in his textbook. Neither do I, though I don't find it particularly unattractive—not by comparison with, say, "Nullstellensatz." Jeff Miller's useful Web site on the earliest known uses of mathematical terms cites Italian geometer Eugenio Beltrami as the culprit, in 1869. Scanning through papers of that date in Beltrami's *Opere Matematiche* (Milano, 1911), I could not find the term. I can't read Italian, though, so my failure should not be taken as dispositive.

139. I have glossed over the fact that, from the mid-19th century on, geometry has embraced complex-number coordinates. This is conceptually hard to get used to at first, which is why I have glossed over it. One consequence, for example, is that if you admit complex numbers into coordinates and coefficients, a line can be perpendicular to itself! (In ordinary Cartesian coordinates, two lines with gradients m_1 and m_2 are perpendicular when $m_1 \times m_2 = -1$. A line with gradient i is therefore perpendicular to itself.) Similarly, teachers of higher algebraic geometry relish the say-*what?* moment when their undergraduate students, fresh from wrestling with the complex-number plane in their analysis course, are introduced to the complex-number *line*. (That is, a one-dimensional space whose coordinates are complex numbers. If you are confused by this—you should be.)

140. Riemann surfaces provided mid-20th-century novelist Aldous Huxley with an item of scenery. Readers of Huxley's novel *Brave New World* will recall that the citizens of the year 632 After Ford amused themselves by playing Riemann-surface tennis.

141. Gustav Roch (1839–1866) studied under Riemann at Göttingen in 1861. He died very young—not quite 27—four months after Riemann himself.

142. Though Stubhaug, at any rate in Richard Daly's translation, occasionally displays a deft literary touch that tickles my fancy. Of one hiking trip in Lie's student days, Stubhaug notes: "[I]t was further along the way that they met the three beautiful, quick-witted alpine milkmaids, who, in Lie's words, were 'free of every type of superfluous reticence.' However, the distance they penetrated into Jotunheimen that summer seems uncertain."

143. Once again I am oversimplifying disgracefully. In fact, as Klein knew, Euclid's actual propositions remain true if you include *dilatations*—that is, uniform enlargements or shrinkages of the entire plane, turning figures into other figures of the same shape but different sizes. I am going to ignore this complication. The reader who wants to explore it is referred to Chapter 5 of H. S. M. Coxeter's 1961 classic textbook *Geometry*.

144. In German, *Erlanger Programm*, the *-en* getting inflected to *-er*. This is often carried over into English, so you will see "Erlanger program." This seems wrong to me; but it is so commonly done, there is no use complaining.

145. In a nutshell, a Lie group is a group of continuous transformations of some general n-dimensional manifold that has important properties of "smoothness." A Lie algebra is an algebra just as I defined the term in my §VS.6: a vector space with a way to multiply vectors. The vector multiplication in a Lie algebra is of a rather peculiar sort, but turns out to be very useful in certain high-calculus applications, and to arise naturally out of Lie groups.

146. Dirk Struik, reviewing Coolidge's *History of Geometrical Methods* (1940).

147. Asked if he was related to these high-class Brookline Coolidges, the 30th president, whose origins were much humbler, replied with the brevity for which he was celebrated: "They say not." In fact, practically all American Coolidges are descended from the five sons of John Coolidge of Watertown, 1604–1691. The president was of the eighth generation after the second son, Simon; the mathematician was of the seventh generation from the fifth son, Jonathan; so president and mathematician were seventh cousins once removed. Julian Coolidge's grandmother was a granddaughter of Thomas Jefferson.

Chapter 14: Algebraic This, Algebraic That

148. The business of two objects being "the same"—topologically equivalent—under properly supervised stretching and squeezing fairly cries out for a nice snappy bit of jargon to encompass it. The usual term of

art is *homeomorphic.* There is a bit more to be said about that, though, and for simplicity's sake in a popular presentation I shall go on saying "topologically equivalent."

149. C_∞ is named "the infinite cyclic group." If you use multiplication as the shorthand way of expressing the group composition rule, C_∞ consists of all powers, positive, negative, and zero, of one element a: ... a^{-3}, a^{-2}, a^{-1}, 1, a, a^2, a^3, ... Since multiplying two powers of a is done by just adding their exponents ($a^2 \times a^5 = a^7$), another instantiation of C_∞ is the group of ordinary integers in $\mathbb{Z}$ with the operation of addition. For this reason, you will sometimes see the fundamental group of the torus given as $\mathbb{Z} \times \mathbb{Z}$ or, more meticulously, since $\mathbb{Z}$ names a ring, not a group, as $\mathbb{Z}^+ \times \mathbb{Z}^+$.

150. Also, sometimes, a three-sphere. This terminology is, though, hard to keep straight in one's mind, at least for nonmathematicians. Does "three-sphere" refer to the two-dimensional surface of an ordinary sphere, curved round and dwelling in three-space? Or to the impossible-to-visualize three-dimensional surface of a hypersphere, curved round and dwelling in four-space? To a mathematician it is the latter, ever since Riemann taught us to think about a manifold—a space—from a vantage point within the manifold itself. To a layperson, however, more used to seeing two-dimensional surfaces surrounded by three-dimensional space, the former is just as plausible.

151. A related theorem, due to topologist Heinz Hopf (1894–1971), and often confused with Brouwer's FPT, assures us that at some point on the Earth's surface at this moment, there is absolutely, though instantaneously, no wind. Or equivalently, imagine a sphere covered with short hair, which you are trying to brush all in one direction. You will fail. No matter how you try, there will always be one (at least) "whorl point" where the hair won't lie down. This has led to the theorem being referred to rather irreverently by generations of math undergraduates as "the cat's anus theorem." (I have bowdlerized slightly.) Thus considered, the theorem states: Every cat must have an anus.

152. Though G. T. Kneebone says a thing that needs to be said here: "Kant's conception of mathematics has long been obsolete, and it would be quite misleading to suggest that there is any close connexion between it and the intuitionist outlook. Nevertheless it is a significant fact that the

intuitionists, like Kant, find the source of mathematical truth in intuition rather than in the intellectual manipulation of abstract concepts" *Mathematical Logic and the Foundations of Mathematics*, p. 249.

153. Which, like the Poincaré conjecture (§14.2) is one of the problems for which the Clay Institute is offering a million-dollar prize. See Keith Devlin's 2002 book, *The Millennium Problems*, for a full account of all seven problems.

154. *Proof that the limit of the sequence is* $\sqrt{2}$: By the rule for forming terms, if some term of the sequence is $\dfrac{a}{b}$, then the following term is

$$\frac{a+2b}{a+b},$$

which is

$$\frac{(a+b)+b}{a+b},$$

which is

$$1+\frac{b}{a+b},$$

which is

$$1+\frac{1}{1+\frac{a}{b}},$$

which is

$$1+\frac{1}{1+(previous\ term)}$$

If the sequence closes in on some limiting number, the terms get closer and closer together, so that *term* and *previous term* are well-nigh equal. So after a few trillion terms, it is well-nigh the case that

$$x = 1 + \frac{1}{1+x}$$

That is, if you apply some elementary algebra, a quadratic equation, whose only positive root is $x = \sqrt{2}$. Q.E.D. This proof is, of course, not rigorous, its principal weakness being that "If" at the start of the second sentence.

155. *Man muss jederzeit an Stelle von „Punkten, Geraden, Ebenen," „Tische, Stühle, Bierseidel" sagen können.* Hilbert is very quotable.

156. Co-authored by S. Cohn-Vossen and published in 1932, two years after Hilbert retired from teaching.

157. Weil, like Lie, had the unhappy distinction of having been arrested as a spy, his mathematical notes and correspondence taken for encrypted communications. That was in Finland, December 1939. Released and sent back to France, he was arrested there for having evaded military service.

Chapter 15: From Universal Arithmetic to Universal Algebra

158. Professor Barry Mazur, himself a skillful and lucid popularizer of math, set out to explain the concept of a *motive* (as in "motivitic") to non-algebraical readers of the November 2004 *Notices of the American Mathematical Society*. His article, which I believe does the job as well as it can be done, begins: "How much of the algebraic topology of a connected finite simplicial complex X is captured by its one-dimensional cohomology?"

159. By the AMS classification number codes, the 13 are: (06) Order, lattices, ordered algebraic structures; (08) General algebraic systems; (12) Field theory and polynomials; (13) Commutative rings and algebras; (14) Algebraic geometry; (15) Linear and multilinear algebra, matrix theory; (16) Associative rings and algebras; (17) Nonassociative rings and algebras; (18) Category theory, homological algebra; (19) K-theory; (20) Group theory and generalizations; (22) Topological groups, Lie groups; and (55) Algebraic topology.

160. Mac Lane argued, with what validity I do not know, that Birkhoff Senior's policy was motivated at least in part by plain patriotism in the jobless 1930s. Mac Lane: "George Birkhoff at Harvard . . . felt that we also ought to pay attention to young Americans, so there were relatively few appointments of [European] refugees at Harvard" (from the book *More Mathematical People*, 1990, Donald J. Albers, Gerald L. Alexanderson, and Constance Reid, Eds.)

161. Professor Swan adds the following interesting historical note: "The homotopy groups were discovered by [Eduard] Cech in 1932, but when he found they were mainly commutative he decided that they were uninteresting, and he withdrew his paper. A few years later, [Witold] Hurewicz rediscovered them, and is usually given the credit."

162. "Pentatope" is the word H. S. M. Coxeter uses for this object in his book *Regular Polytopes* (Chapter 7). I have not seen the word elsewhere, don't know how current it is, and do not think it would survive a challenge in Scrabble. As I have shown it, of course, the wire-frame pentatope has been projected down from four dimensions into two, so the diagram is very inadequate.

163. A procedure related to the notion of *duality* that crops up all over geometry. The classic "Platonic solids" of three-dimensional geometry illustrate duality. A cube (8 vertices, 12 edges, 6 faces) is dual to an octahedron (6 vertices, 12 edges, 8 faces); a dodecahedron (20 vertices, 30 edges, 12 faces) is dual to an icosahedron (12 vertices, 30 edges, 20 faces); a tetrahedron (4 vertices, 6 edges, 4 faces) is dual to itself. I ought to say, by the way, in the interest of historical veracity, that it was Emmy Noether who pointed out the advantage of focusing on the group properties here. Earlier workers had described the homology groups in somewhat different language.

164. The term "universal algebra" has an interesting history, going back at least to the title of an 1898 book by Alfred North Whitehead, the British mathematician philosopher of *Principia Mathematica* co-fame (with Bertrand Russell). Emmy Noether used it, too. My own usage here, though, is only casual and suggestive and is not intended to be precisely congruent with Whitehead's usage, or Noether's, or anyone else's.

165. Category theory's only appearance in popular culture, so far as I know, was in the 2001 movie *A Beautiful Mind*. In one scene a student says to John Nash: "Galois extensions are really the same as covering spaces!" Then the student, who is eating a sandwich, mumbles something like: "... functor ... two categories ...." The implication seems to be that Galois extensions (see my primer on fields) and covering spaces (a topological concept) are two categories that can be mapped one to the other by a functor—quite a penetrating insight.

166. Allyn Jackson quotes some revealing remarks about this by Justine Bumby, with whom Grothendieck was living at the time: "His students in mathematics had been very serious, and they were very disciplined, very hard-working people. . . . In the counterculture he was meeting people who would loaf around all day listening to music."

167. The Grothendieck Biography Project, at *www.fermentmagazine.org/home5.html.*

168. Both the Lorentz group and the one used by Gell-Mann to organize the hadrons—technically known as the special unitary group of order 3—can be modeled by families of matrices, though the entries in the matrices are complex numbers.

169. The precise definition, just for the record, is "A Riemannian manifold admitting parallel spinors with respect to a metric connection having totally skew-symmetric torsion."

170. Yau, winner of both a Fields Medal and a Crafoord Prize, was a "son of the revolution," born in April 1949 in Guangdong Province, mainland China. Following the great famine and disorders of the early 1960s, his family moved to Hong Kong, and he got his early mathematical education there. He is currently a professor of mathematics at Harvard.

171. Printed up as "Mathematics in the 20th Century" in the *American Mathematical Monthly*, 108(7).

172. The sense here is that Newton was the absolute-space man, while Leibniz was more inclined to the view that, as the old ditty explains:

    Space
    Is what stops everything from being in the same place.

173. It is a common misconception that Einstein banished all absolutes from physics and hurled us into a world of relativism. In fact, he did nothing of the sort. Einstein was as much of an "absolutist" as Newton. What he banished was absolute space and absolute time, replacing both with *absolute space-time.* Any good popular book on modern physics should make the point clear. Einstein's close friend Kurt Gödel was, by the way, a strict Platonist: The two pals were yin and yin.

PICTURE CREDITS

Unattributed pictures and pictures for which I have identified only a documentary source are those I believe to be in the public domain, or for which I have been unable to locate any holder of rights.

Nobody is keener on the laws of copyright than an author, and I have done my honest best to track down the proper parties for permission to reproduce photographs and pictures in this book. Locating those parties is, however, not easy. It is especially difficult in the case of pictures one finds on the internet, where nobody properly credits anything.

Should any person or institution feel that his, her, or its rights have been overlooked by my use of pictorial material in this book, I can only ask that he, she, or it get in touch with me through the publisher, so that I can make appropriate restitution, which I shall gladly and speedily do.

Neugebauer: Courtesy of the John Hay Library at Brown University, Providence, Rhode Island.

Hypatia: Painting by Charles William Mitchell (1854–1903), reproduced by permission of Tyne & Wear Museums, Newcastle upon Tyne, England.

Cardano: From the frontispiece of Cardano's *The Great Art, or The Rules of Algebra*. I have actually taken it from the M.I.T. Press edition of that work, translated and edited by T. Richard Witmer (Cambridge, Massachusetts, 1968).

Viète: From the frontispiece of *François Viète: Opera Mathematica*, recognita Francisci A. Schooten; Georg Olms Verlag; Hildesheim (New York, 1970).

Descartes: An engraving by an artist unknown to me, taken from Franz Hals's 1649 painting, which is in the Louvre, Paris.

Newton: An 1868 engraving by Thomas Oldham Barlow from the 1689 portrait by Godfrey Kneller, which is in the Wellcome Library, London.

Leibniz: Engraving from a painting in the Uffizi Gallery, Florence.

Ruffini: Taken from the frontispiece of *Opere Matematiche di Paolo Ruffini*, Vol. 1, Tipografia Matematica di Palermo, Italy (1915).

Cauchy: From the portrait by J. Roller (ca. 1840), by permission of École Nationale des Ponts et Chaussées, Champs-sur-Marnes, France.

Abel: From *Niels-Henrik Abel: Tableau de Sa Vie et Son Action Scientifique* by C.-A. Bjerknes; Gauthier-Villars (Paris, 1885).

Galois: "Portrait d'Évariste Galois a quinze ans," from the *Annales de l'École Normale Supérieure*, 3c série, Tome XIII (Paris, 1896).

Sylow: University of Oslo library, Norway.

Jordan: Taken from *Oeuvres de Camille Jordan*, Vol. 1, edited by J. Dieudonne; Gauthier-Villars & Cie., Editeur-Imprimeur-Libraire (Paris, 1961).

Hamilton: Portrait by Sarah Purser (from a photograph); courtesy of the library of the Royal Irish Academy, Dublin.

Grassmann: Taken from *Hermann Grassmanns Gesammelte Mathematische und Physikalische Werke*, Chelsea Publishing Company (Bronx, New York, 1969). By permission, American Mathematical Society.

Riemann: Courtesy of the Staatsbibliothek zu Berlin, Preussischer Kulturbesitz.

Abbott: Courtesy of City of London School.

Plücker: From *Julius Plückers Gesammelte Mathematische Abhandlungen*, edited by A. Schoenflies, Druck und Verlag von B.G. Teubner (Leipzig, Germany, 1895).

Lie: Portrait by Joachim Frich, courtesy of the University of Oslo, Norway.

Klein, Dedekind, Hilbert, Noether: Courtesy of Niedersächsische Staats- und Universitätsbibliothek, Göttingen, Germany; Abteilung für Handschriften und seltene Drucke.

Lefschetz: By permission of the Department of Rare Books and Special Collections, Princeton University Library, New Jersey.

Zariski, Grothendieck: Courtesy of the Archives of the Mathematisches Forschungsinstitut Oberwolfach, Germany.

Mac Lane: University of Chicago Library.

The Calabi–Yau illustration (Figure 15-3) was created by Jean-François Colonna of the Centre de Mathématiques Appliquées at the École Polytechnique in Paris. It is reproduced here with his permission.

INDEX